CW01213470

Winning Without Fighting

WINNING WITHOUT FIGHTING

Irregular Warfare and Strategic
Competition in the 21st Century

Rebecca Patterson, Susan Bryant,
Ken Gleiman, and Mark Troutman

Rapid Communications in Conflict and Security Series
General Editor: Thomas G. Mahnken
Founding Editor: Geoffrey R.H. Burn

CAMBRIA
PRESS

Amherst, New York

Copyright 2024 Cambria Press.

All rights reserved.

Library of Congress Cataloging-in-Publication Data

Names: Patterson, Rebecca, 1974- author. | Bryant, Susan F., author. | Gleiman, Ken, author. | Troutman, Mark, author.

Title: Winning without fighting : irregular warfare and strategic competition in the 21st century / Rebecca Patterson, Susan Bryant, Ken Gleiman, and Mark Troutman.

Other titles: Irregular warfare and strategic competition in the 21st century

Description: [Amherst, New York] : Cambria Press, [2024] | Series: Rapid communications in conflict and security series | Includes bibliographical references and index. | Summary: "This book offers a framework for an American grand strategy with a corresponding approach for visualizing progress towards its achievement. The authors contend that instead of conventional military conflict, irregular warfare (IW)--"activities short of conventional and nuclear warfare that are designed to expand a country's influence and legitimacy, as well as to weaken its adversaries"--along with transnational threats, characterize today's competitive landscape. Thus, in addition to military capabilities, U.S. grand strategy must extend beyond the purview of the DOD. Given that this is an era of crises as well as of competition, any effective strategy must also include ways to preserve and bolster national resilience"-- Provided by publisher.

Identifiers: LCCN 2024014856 (print) | LCCN 2024014857 (ebook) | ISBN 9781638571940 (library binding) | ISBN 9781638573210 (paperback) | ISBN 9781638573234 (pdf) | ISBN 9781638573241 (epub)

Subjects: LCSH: Irregular warfare--United States. | National security--United States. | Strategic culture--United States.

Classification: LCC U167.5.I8 P37 2024 (print) | LCC U167.5.I8 (ebook) | DDC 355.02/18--dc23/eng/20240415

LC record available at https://lccn.loc.gov/2024014856

LC ebook record available at https://lccn.loc.gov/2024014857

*Dedicated to Anna Matilde Bassoli,
Alison O'Neil, and Christian Trotti.
Without your incredible assistance,
this book would not have been possible.*

Table of Contents

List of Figures ix

Glossary xi

Foreword xvii

Chapter 1: Introduction 1

Chapter 2: American Strategic Culture 21

Chapter 3: Competitor Approaches to Irregular Warfare 43

Chapter 4: The Past as Potential Prologue: Political Warfare
during the Cold War 67

Chapter 5: Tools of Military Statecraft 89

Chapter 6: Tools of Economic Statecraft 113

Chapter 7: Tools of Information Statecraft 141

Chapter 8: Tools of Resilience 167

Chapter 9: Measuring Success 193

Chapter 10: Conclusion: An American Grand Strategy for
Winning Without Fighting 213

Bibliography 233

Index 311

About the Authors 329

Cambria Rapid Communications in Conflict and Security
Series 331

List of Figures

Figure 1: Core Terminology 7

Figure 2: DOD Definitions of IW 10

Figure 3: Definitions of IW Objectives 13

Figure 4: Competitor Strategic Culture and Use of IW Tools 49

Figure 5: Typology of Military Statecraft Tools 95

Figure 6: Typology of Economic Statecraft Tools 116

Figure 7: Typology of Information Statecraft Tools 153

Figure 8: Typology of Resilience Tools 175

Figure 9: Irregular Warfare Dashboard 205

Glossary

active measures – A term used to describe political warfare conducted by the Soviet Union and Russia generally referring to covert operations ranging from disinformation campaigns to staging insurrections.

anti-access/area denial (A2/AD) – Military capability that prevents or restricts the ability of adversaries from entering or operating in specific geographic areas.

arms control – Diplomatic and strategic efforts to limit, regulate, or reduce production, deployment, proliferation, and use of military weapons and technologies.

collective resilience – The collective capacity of allied states to withstand, fight through, and quickly recover from disruption, either military or non-military, caused by adversaries and global challenges.

community resilience – A community's ability to respond to, withstand, and recover from adverse situations.

compellence – Efforts employed to change the behavior of another actor or state by coercing them to undertake or cease a particular course of action.

conditionalism – A belief that economic factors are subordinate to national security imperatives but that economic liberalization can lead to increased power, influence, and legitimacy.

containment – A key element of US foreign policy during the Cold War that aimed to limit the expansion of communist-led governments.

conventional forces – Military assets and personnel that use traditional, non-nuclear, and non-chemical weaponry and tactics.

defense innovation – The process of developing and incorporating new technologies, strategies, and concepts to enhance military capabilities, effectiveness, and security.

deterrence – Efforts aimed at preventing or discouraging an adversary from taking a specific course of action by signaling that the costs, risks, or consequences of such actions outweigh potential benefits.

disinformation – Deliberately false information spread with deceptive intent.

economic liberalism – A school of thought that emphasizes the importance of individual freedom, property rights, and free markets as the key drivers of economic prosperity and progress.

force posture – The strategic positioning, deployment, and configuration of a nation's military forces and capabilities to achieve specific defense and security objectives.

freedom of navigation operation (FONOPS)– A military operation, often in disputed air or maritime areas, to assert and uphold in the legitimacy of international waters and airspace.

Goodhart's law – The sociological analog of Heisenberg's Uncertainty Principle. It states that when a measure becomes the target, it often ceases to be a good measure.

gray zone – The ambiguous environment that exists between peace and war.

Hamiltonianism – A US political subculture characterized by a strong national government and powerful alliances.

hybrid warfare – A type of warfare that uses regular and irregular tactics and/or actors.

individual resilience – A person's capacity for adapting psychologically, emotionally, and physically reasonably well and without lasting detriment to self, relationships, or personal development in the face of adversity, threat, or challenge.

influence – The ability to affect others' beliefs, that is, their knowledge or opinions, either about what is or what ought to be the case.

influence operations – The coordinated, integrated, and synchronized application of national diplomatic, informational, military, economic, and other capabilities to foster attitudes, behaviors, or decisions by foreign target audiences that further a state's interests.

information – Data that has been processed, organized, or structured in a meaningful way to convey knowledge, facts, or instructions.

information operations – The integrated employment of information-related capabilities to influence, disrupt, corrupt, or usurp the decision-making of adversaries while protecting one's own.

information statecraft – The means by which a state uses information and narrative tools to achieve its strategic objectives.

information warfare – The use of information and communication technologies and platforms to manipulate and influence public opinion, disrupt societal harmony, spread information or disinformation, and undermine the stability and security of a state.

international resilience – The ability of the international community, comprised of states, international organizations, alliances, and the private sector, to withstand, adapt to, and recover from shocks.

irregular warfare – Activities short of conventional and nuclear warfare that are designed to expand a country's power, influence, and legitimacy, as well as to weaken its adversaries.

Jacksonianism – A US political subculture mainly concerned with physical security and economic prosperity.

Jeffersonianism – A US political subculture mainly concerned with safeguarding democracy at home.

legitimacy – A perception or assumption that the actions or standing of an entity are desirable, proper, or appropriate within socially constructed norms, values, and beliefs.

measure of effectiveness (MOEs) – An indicator used to measure a current system state, with change indicated by comparing multiple observations over time associated with achieving a particular outcome.

measure of performance (MOPs) – An indicator used to measure task accomplishment associated with achieving a particular output.

mercantilism – A school of thought skeptical of the role of economic factors in persuading states to change behavior associated with government intervention in the economy, protectionism, and a focus on achieving a positive balance of trade.

military exercises – Activities conducted by armed forces to practice and test their capabilities, strategies, tactics, and procedures.

military statecraft – A state's use of military means to achieve foreign policy ends without the use of force.

misinformation – False information shared without the intent to deceive.

national resilience – A state's ability to withstand, adapt to, and recover from a wide range of challenges, crises, and disruptions, both internal and external, while maintaining its core functions, values, and security.

political culture – The mix of attitudes, beliefs, emotions, and values of society that relate to the political system and political issues, which may be held implicitly.

political warfare – The employment of all the means at a nation's command, short of war, to achieve its national objectives.

power – The ability to attain desired outcomes by coercing others.

propaganda – The systematic and strategic dissemination of information, ideas, or opinions with the aim of influencing and manipulating the beliefs, emotions, attitudes, and behaviors of individuals or groups for political, ideological, or persuasive purposes.

resilience – The ability to withstand, adapt to, and recover from a wide range of challenges, crises, and disruptions while maintaining core functions, values, and security.

security cooperation – Interactions with foreign security establishments to build relationships that promote specific security interests, develop allied and partner nation military and security capabilities, and provide peacetime and contingency access.

strategic competition – A persistent and long-term struggle occurring between two or more adversaries seeking to pursue incompatible interests without necessarily engaging in armed conflict with each other.

strategic culture – The set of ideas and beliefs that influence how a nation perceives threats, formulates its strategic objectives, and makes decisions about the use of tools of statecraft.

three warfares – A Chinese irregular warfare framework that consists of "public opinion warfare" to influence the perceptions of domestic and foreign publics; "psychological warfare" to build or exert influence over foreign elected leaders, opposition groups, or military leaders, affecting the execution of their policies vis-à-vis China; and "legal warfare" to bolster legitimacy by constructing legal justifications for Chinese actions and promoting alternative interpretations of international law.

Wilsonianism – A US political subculture characterized by a moral obligation and national interest to spread democracy.

Foreword

"To fight and win our nation's wars"—over thirty years in the Department of Defense (DOD) and almost another decade in positions partnered with it, I have taken part in the debate over what this means. Which kinds of "wars" will we actually "fight"? And what are the most effective tools with which to do that? From the Cold War to the Peace Operations of the 1990s, to two decades of counterterrorism and counterinsurgency, to the present era of strategic competition, one thing has been clear to me: "irregular warfare" is anything but irregular.

Currently, the name of the game is deterring major power war, and this indeed demands a modern and ready conventional and nuclear force of sufficient size. The US has some catching up to do. But the requirements of the more likely conflicts—countering gray-zone coercion of many forms, proxy wars in support of like-minded governments and populations, and fighting for influence with information and a range of other tools—are not necessarily inherent in what is called "traditional military activities." They demand more thought and investment, more nuance and patience.

Given this context, a couple of dynamics are at play that make *Winning Without Fighting: Irregular Warfare and Strategic Competition in the 21st Century* timely and important.

First, the 2022 National Defense Strategy elbowed "strategic competition" slightly off to the side to offer center stage to "integrated deterrence." Without prejudice to the appropriateness of this shift, the latter remains a critical target for study, capability development, and renewed conversation among the many government organizations. Indeed, the risk of the United States being outcompeted within the world economy, the diplomatic realm, and the information environment likely outweighs the risk of a major power war. The Joint Staff followed suit, prioritizing the development of the *Joint Warfighting Concept* yet keeping the excellent *Joint Concept for Competing* alive in recognition that "winning" really means gaining and maintaining advantage indefinitely.

Second, "irregular warfare" remains a much-maligned and misunderstood concept in certain corners of the national security apparatus. Some believe that irregular warfare will lead to conventional warfare—that it is simply a precursor to a big fight—and reject the notion of robust collaboration with the military. Others, harkening back to the recent counterinsurgency past, fear the military will once again prioritize "nontraditional" roles, invading spaces that are appropriately civilian, as well as detracting from the DOD's primary mission: to fight and win our nation's wars. In fact, a legislative proposal suggesting that the definition of traditional military activities include activities associated with irregular warfare was roundly rejected. This is unfortunate, given the capabilities and activities included under the "irregular warfare" rubric. *Winning Without Fighting* helps to clarify some of those misconceptions.

At the same time, Congress recognizes the critical need to better understand opportunities and challenges in the "gray zone" and authorized the creation of the Irregular Warfare Functional Center (IWFC) with the Mac Thornberry National Defense Authorization Act of 2021. According to the legislation, such a center could "enhance and sustain focus on, and advance knowledge and understanding of, matters of irregular warfare." It is likely that this book and the authors themselves will be of much use to the IWFC in the learning it imparts.

Foreword

I have known the authors for years. They embody the model of the scholar-practitioner. Each of them has deep, personal experience with the military and irregular warfare (IW) as well as significant interagency and international experience. They have all steadily worked on, led teams toward, and taught these supposed "non-traditional" military activities and effects. The various chapters assembled here reflect years of study and work involving not only the DOD but many other academic and government organizations. The result is a thoughtful and practically grounded book useful to anyone, whether in uniform or not, confronting the problem sets related to irregular warfare and strategic competition.

As Deputy Assistant Secretary of State in the Bureau of Conflict and Stabilization Operations, I helped launch the Stabilization Assistance Review, which aimed to capture the principles and roles of various agencies in stabilization efforts as well as the significant lessons of the whole-of-government operations meant to assist other states in stabilizing across their societies. While the geopolitical context the US government has chosen to focus on appears vastly different— less concerned about the internal affairs of other states and more concerned once again with inter-state relations—one thing remains the same: the Department of Defense will solve none of our security challenges alone.

As Director of Strategy, Policy, and Plans at US Special Operations Command, it is clear to me that the name of the game is "irregular"— asymmetric, indirect, subtle, precise, agile, partnered. Above all, IW demands partnerships and informing, supporting, and enabling others. While pursuing and protecting our national interest has always been a team sport—an infinite, complex game—at this moment in history, winning demands a total greater than the sum of the individual agency parts.

Russia's 2022 full-scale invasion of Ukraine has directly undermined the rules-based international order, though, for years, they have used hybrid tactics to coerce West-leaning former Soviet satellites to do their bidding. In addition, Russia's adroitness with (undemocratic) tools of information warfare has amplified divisions in the polity of more than one country. Finally, Russia's proxies—the Wagner Group, in particular, have contributed to the destabilization of the countries in which they operate.

Hamas's brazen attacks on innocent civilians in southern Israel, Israel's punitive response and resultant civilian casualties, the activation of the "Axis of Resistance" led by Iran, the position in which Jordan and the Gulf states find themselves, as well as the domestic turmoil in the US over support for Israel versus the Palestinian cause, have destabilized the Middle East so much that the United States' directed focus on the "pacing challenge" of China is at risk. Irregular warfare options provide the highest non-escalatory potential to manage Iran, assist Israel within the bounds of our democratic values, and message a middle ground way ahead.

The challenge of China is real. The need for a nuclear and conventional military—ready, modern, and of sufficient size—is, therefore, real. While deterrence works and is rightfully the cornerstone of our defense strategy, it is not designed to be sufficiently agile to compete day to day. We must deter in other modalities—diplomatic, economic, and information—and combine these capabilities with military ones, which are not necessarily lesser included subsets of the larger deterrence force. In fact, a focus on combining means and associated authorities indicates the need for a different articulation of the ways and ends as well as a more nuanced national grand strategy that compels non-DOD-centric resourcing.

Ultimately, this book could nudge the US national security machine in that direction. Hopefully, it will encourage national security professionals, particularly military leaders, to open their minds to the necessity of partnering when they are not leading and robustly participating in the learning and operational activities of other agencies and elements of society. There is no better way for the DOD to contribute to winning.

—Brigadier General Kim Field, US Army (ret.),
Director for Strategy, Plans, and Policy,
United States Special Operations Command

Winning Without Fighting

CHAPTER 1

INTRODUCTION

We face two strategic challenges. The first is that the post–Cold War era is definitely over, and a competition is underway between the major powers to shape what comes next...The second is that while this competition is underway, people all over the world are struggling to cope with the effects of shared challenges that cross borders—whether it is climate change, food insecurity, or communicable diseaseThese shared challenges are not marginal issues that are secondary to geopolitics. They are at the very core of our national and international security. [1]
The National Security Strategy of the United States, 2022

INTRODUCTION

Since the end of the Cold War, US national security policy has suffered strategic drift.[2] A country that once boasted of major foreign policy successes—vanquishing Germany and Japan on the battlefield, establishing a new postwar international order, and capably defending that order from the Soviet challenge—has since struggled to define its purpose in the world. From the short-lived "unipolar moment" in the 1990s to the Global War on Terror and the conflicts in Iraq and Afghanistan, America has lacked a guiding grand strategy for leveraging its impressive means to achieve major ends and defend vital interests. As a result, the United States has lost wars, spent an inordinate sum, and stumbled from one priority to another.

Today, however, the most dangerous revisionist challengers to the US and its allies and partners are clear for the first time since the end of the Cold War. China's intention and capacity to "reshape the international order"[3] through various means is the most transformative geopolitical development of the twenty-first century to date.[4] Concurrently, Russia remains a persistent threat. The Kremlin has attacked American democracy through election-related disinformation campaigns and challenged the rules-based global order with its 2014 annexation of Crimea and its 2022 invasion of Ukraine. These challenges have focused US attention. Several administrations[5] have successively prioritized the reemergence of long-term, strategic competition as the central threat to American prosperity and security. This newfound consensus should spark hope that a coherent grand strategy is once again possible in Washington.

The Central Argument

Although the importance of competition with China and Russia has gained considerable traction—both inside the Beltway and in academia—there is far less consensus about the gravity of today's transnational threats. The COVID-19 pandemic is one of many shocks that have plagued the earth in the recent past—from the global financial crises and instability in the Middle East to the rise of right-wing extremism to countless natural disasters—the global environment has been extremely turbulent. In fact, Mathilde Tomine Eriksdatter Giske refers to the disorder as an "age of crises,"[6] a term which this book adopts.

That moniker is far from hyperbolic for several reasons. First, no country is safe from the accelerating effects of climate change. Extreme weather patterns and rising sea levels threaten to erode and ultimately destroy critical infrastructure—both civil and military. Key American military bases at home and abroad, necessary for sustaining its global power-projection capabilities, will be rendered unusable or even underwater in a matter of decades—yet little has been done to increase their resilience.[7]

Moreover, climate change increases public health risks. In 2023, distant Canadian wildfires spread effects over cities along the Eastern Seaboard, sending tens of thousands of people indoors to avoid exposure to harmful air quality. Health Canada estimates that between 2013 and 2018, decreases in air quality from more frequent wildfires accounted for up to 240 annual and 2500 long-term premature deaths. The cost of related acute and chronic health impacts may exceed $20 billion.[8] As average global temperatures rise, natural disasters grow more likely, potent, and costly.

Perhaps just as virulently, the diffusion of advanced technologies such as artificial intelligence (AI), autonomous technology, and cyber technology are imbuing ever smaller non-state, terrorist, and criminal actors with more powerful capabilities. In the twenty-first century, data and information are power. Thus, additional safeguards are required to defend against data theft, disinformation and misinformation, and other emerging technological challenges.

Although all states must contend with these transnational threats, their performance will not be even. The states that can better address the effects of shocks and the diffusion of technology will secure advantageous positions. Given that the current threat environment can be characterized simultaneously as an arena of strategic competition and an era of crises—the mandate of national security must expand to address this nexus.

In the face of these challenges, there are significant impediments to formulating a coherent US approach. First, American policymakers often do not understand the nature of today's geopolitical threats and the subsequent ineffectiveness of their preference for using conventional military means or over-reliance on sanctions over other tools of statecraft. Unfortunately, this construct is ill-suited to either strategic competition or an era of crises. Though China and Russia seek to contest the ability of the US to legitimately wield power and influence and foster an alternative international order that promotes values like sovereignty over human rights,[9] they are unlikely to initiate World War III to achieve those outcomes.[10]

Modern war is simply too costly, especially when fought against a technologically advanced military. Instead, US competitors employ political, economic, and information means much more frequently than military force, operating below the threshold of what the US would consider war. For example, international influence is often bought through economic assistance, and disinformation is used to diminish American legitimacy internationally and foment discord domestically. Against such a diverse playbook, US military power—while still necessary to deter more aggressive gambits—is insufficient. And it is largely useless when confronting transnational threats like climate change and disease. Thus, in a decision-making environment defined by limited resources and boundless, multidimensional threats, policymakers cannot afford to place all of their proverbial eggs in one basket. Military power is not a panacea.

Although scholars and policymakers continue to acknowledge that interstate rivalries will not likely escalate to full-scale combat,[11] that America cannot win or outcompete by fighting a conventional conflict,[12] and that transnational threats will require an increasingly holistic approach,[13] the *de facto* leading agent of US foreign policy remains the Department of Defense (DOD). The resulting circumscribed and highly militarized frame renders American grand strategy myopic, if not absent. Continuing to view grand strategy through the lens of conventional military capability is a recipe for failure in the coming decades, much as it was at the beginning of this century.

Carl von Clausewitz notes,

> The supreme, the most far-reaching act of judgment that the statesman and commander have to make is to establish by that test

the kind of war on which they are embarking; neither mistaking it for, nor trying to turn it into, something that is alien to its nature.[14]

US policymakers have failed to heed this admonition. To prevail in an era characterized by both strategic competition and unrelenting transnational threats, US policymakers must understand that the irregular and nonmilitary facets of today's competition are paramount. To succeed, they must, therefore, frame and execute a comprehensive, whole-of-society grand strategy that harnesses all available means. Until that occurs, American power, influence, and legitimacy will continue to erode.[15]

Further, due to deeply rooted beliefs that characterize American strategic culture, US policymakers often perceive a discernable line exists separating war from peace. Thus, in preparing for the next large-scale conflict, Washington regularly overlooks critical tools like economic coercion and information operations in favor of more military-centric conceptions of war and warfare.[16] Western policymakers often actively avoid the term "warfare" in describing non-kinetic activities because they believe military operations bear their own unique logic.[17] The term "warfare" to describe non-military activities remains unpalatable to many US allies and partners.[18]

Although the US, along with its partners and allies, perceive non-kinetic tools of statecraft as activities to avoid war, China and Russia consider them as forms of war and therefore invest more heavily in them. Failure to perceive these activities as warfare and central to strategic competition impedes America's ability to compete geopolitically. Similarly, the lack of investment in activities not directly connected to conventional conflict, such as hardening critical infrastructure, has hindered American attempts to combat transnational challenges.

Lastly, divergent and overlapping definitions significantly impede a clear understanding of contemporary threats. Many words have been used to describe both the current security environment and the activities that competitors have undertaken to gain relative advantage within it. Terms such as "political warfare," "hybrid warfare," "gray-zone warfare," and "irregular warfare" attempt to ascribe greater importance to these activities, but they are often ambiguous and yield debates that are more semantic than substantive. Consequently, it is often unclear what today's geopolitical competition is about, where it is waged, and how it is conducted.

This book's purpose is to fill the gaps left by these impediments and offer a framework for an American grand strategy with a corresponding approach for visualizing progress towards its achievement. Rather than conventional military conflict, irregular warfare (IW)—"activities short of conventional and nuclear warfare that are designed to expand a country's influence and legitimacy, as well as to weaken its adversaries"[19]—along with transnational threats, characterize today's competitive landscape. Thus, besides military capabilities, US grand strategy must extend beyond the purview of the DOD. Given that this is an era of crises and of competition, any effective strategy must also include ways to preserve and bolster national resilience.

In many respects, this approach is like those the US executed successfully during the Cold War. To succeed today, the US must distribute investments across multiple elements of national power. This will allow for a more holistic and cost-effective development and employment of the capabilities required to hedge against this plethora of challenges. To do this, the US must pursue not only hard power but also invest in soft power, influence, and legitimacy. This combination will cumulatively ensure the ability to compete geopolitically.

This reorientation of priorities will demand uncomfortable tradeoffs. It may require the US to spend more or to accept increased conventional military risk to address the increasingly untenable threats in the economic, information, and resilience spheres. This approach is more appropriate for an era in which the most likely threat is not a subset of the most dangerous one (i.e., a world war). Rather, challenges that fall outside of the conventional military sphere—IW, climate change, and black swan events like pandemics—pose the most pressing threats to both the US and its allies. In fact, the world is already experiencing several such crises simultaneously across the domains of climate, health, and IW.

This chapter now turns to the relevant lexicon, introducing the existing definitional debate over terminology. We aim to provide clarity and help drive consensus toward a clearer and more coherent vocabulary and have, therefore, chosen particular terms because they each serve a distinct purpose. They also comport with the definitions advanced by the 2022 National Security Strategy (NSS) and the National Defense Strategy (NDS)—thereby ensuring that they are useful and accessible to policymakers.

CORE TERMINOLOGY

Recognizing that no term is without weakness, the book will use the following terms:

Figure 1. Core Terminology.

TERM	DESCRIPTION
Strategic competition	The activities that major powers undertake to improve their relative position in the international system
Era of crises	The cascading impact of transnational and international threats, including adversary IW as well as climate change and pandemic disease
Gray zone conflict	The purposeful actions that exceed the bounds of normal competition but remain below the perceived threshold of conventional warfare
Irregular warfare	The activities that US competitors undertake and those the US must muster more effectively to succeed in strategic competition
Power, influence and legitimacy	The objectives of strategic competition

The deepening rivalry between the US and a range of countries with conflicting interests has been hailed as a new era of strategic competition. But what is strategic competition? Though first mentioned in the 2018 NDS, the Trump administration largely ignored the term in favor of Great Power Competition (GPC). Both describe the growing tensions among the US, China, and Russia.[20] The Biden administration subsequently adopted strategic competition as a replacement to GPC, defining it as the long-term struggle with "powers that layer authoritarian governance with a revisionist foreign policy."[21]

There are many critiques of this choice. First, the word "strategic" means both everything and nothing and, therefore, does not clarify the nature of this competition.[22] Second, and more importantly, competition is neither a "means nor an end itself,"[23] and "the decision to compete with another great power should always be over something specific."[24] Thus, the term fails to acknowledge the methods or aims of this competition. Despite these flaws, "strategic competition" drives the formulation, organization, and execution of US national security policy. For this reason, its broad acceptance is useful in building consensus despite its weaknesses. This book, therefore, adopts the definition of strategic competition from the 2023 *Joint Concept for Competing*:

> [S]trategic competition is a persistent and long-term struggle that occurs between two or more adversaries seeking to pursue incompatible interests without necessarily engaging in armed conflict with each other. The normal and peaceful competition among allies, strategic partners, and other international actors who are not potentially hostile is outside the scope of this concept.[25]

We have chosen this definition because it acknowledges that competition itself is not problematic and omits competition among non-hostile actors. It also highlights the persistent and long-term nature of the struggle, which necessitates a holistic approach. At its core, the competition is about legitimacy, power, and influence, which will largely unfold in the political, economic, and information domains and not through military operations or activities alone.[26] Finally, at the core of strategic competition lie incompatible adversary interests, which drive their actions and activities. We will use other terms to fill in the gaps left by this definition, including where, how, and why this competition occurs.

Today, the competition primarily occurs in the gray zone between war and peace. Key characteristics of gray-zone conflict include the pursuit of political objectives through integrated, cohesive, and simultaneous campaigns; the use of non-kinetic, and especially covert, tools; and the avoidance of deliberate escalation, instead relying on gradualism to achieve results.[27] According to Armin Krishnan, the primary rationale for such an approach is for "major powers to engage in long-term campaigns that seek to weaken a usually more powerful adversary while deliberately avoiding a military response."[28] The gray zone is the most useful construct to describe the ambiguous environment that exists between peace and war, as it captures its fluid and dynamic nature and the ways in which the levers of national power consistently cross such thresholds. Furthermore, it fills the void that US adversaries exploit—the flawed American presumption that there is a distinct line between war and peace. Thus, the term "gray-zone conflict" pushes policymakers to navigate an environment where US competitors consistently employ diverse and dynamic tools that keep the international system on the edge of conventional war.

The third term central to this book is irregular warfare (IW), which describes the often non-kinetic means by which states compete in the gray zone. One major problem with this term is that a binary categorization of threats and tools as regular or irregular is not helpful. Conflicts are rarely strictly one or the other, and thus, the term can obscure the spectrum of competitive activities.[29] The word "irregular" incorrectly implies that these non-military activities are less frequent than regular, conventional military campaigns, exacerbating the already inferior place of IW in the hierarchy of DOD importance. By treating these activities as *warfare* (as do China and Russia), the term IW reinforces the idea that the DOD should lead.

But perhaps the most concerning development is the term's evolving definition in the literature. Scholars and practitioners have attempted to define IW in a variety of ways, initially casting IW as a military-centric approach to smaller-scale conflicts involving specialized forces against non-state actors[30] or certain types of activities outside the scope of conventional or traditional warfare like counterinsurgency,[31] counterterrorism,[32] foreign internal defense (FID), and stability operations.[33] Recently, some have considered IW more holistically, with some authors defining it as a strategy of using influence and coercion to build or erode legitimacy in the eyes of a relevant population for the purpose of power or advantage.[34] This could include a diverse set of kinetic tools (such as proxy war) and non-kinetic tools,[35] such as "information operations (including psychological operations and propaganda), cyber operations, support to state and nonstate partners, covert action, espionage, and economic coercion."[36] This has given rise to two discrete views on IW: those that believe IW must include violence and use of force and those that view IW as actions taken below the threshold of conventional warfare—activities that do not necessitate force or violence.

Figure 2. DOD Definitions of IW.

YEAR	DOD DEFINITION OF IRREGULAR WARFARE
2008-2023	A violent struggle among state and non-state actors for legitimacy and influence over the relevant population(s).
2023-Present	A form of warfare where states and non-state actors campaign to assure or coerce states or other groups through indirect, non-attributable, or asymmetric activities.

Between 2008 and 2023, the DOD insisted that IW had to involve violence but still agreed with many policymakers who gravitated toward defining IW by thresholds and/or strategies aimed at influence and legitimacy for political power.[37] In 2023, however, DOD redefined IW keeping the focus on violence (through coercion) but eliminated the notion that IW was about a different strategy or theory of victory.[38] The new DOD definition took us backwards..

The sole focus on violence as a necessary condition for IW is not appropriate. A holistic conception of IW clarifies the spectrum of activities available for strategic competition. Therefore, IW should be prioritized over other terms to describe the ways and means by which major powers compete today. As such, Seth G. Jones's definition of IW as "activities short of conventional and nuclear warfare that are designed to expand a country's influence and legitimacy, as well as to weaken its adversaries" is appropriate.[39] Violence and coercion have their place but are unnecessary for an action to be part of an IW campaign. Even passive means of expanding power, influence, and legitimacy, such as bolstering resilience to climate change and cyberattacks, can help accrue advantages over competitors. Although this definition broadens the scope of activities that can be considered IW,[40] it is helpful since it may drive a more comprehensive approach to countering China and Russia.

Other terms in use are nearly synonymous with our preferred term and definition of IW. George F. Kennan, the founder of the Cold War containment strategy, defines political warfare as "the employment of all the means at a nation's command, short of war, to achieve its national objectives."[41] Due to concerns about the word "warfare,"[42] some scholars prefer other terms like "competitive statecraft," which Ryan Shaw defines as "the integration and synchronization of all instruments of national power during both peace and war to secure the nation's interests by advancing its values."[43] Thus, Jones's definition of IW, Kennan's definition of political warfare, and Shaw's definition of competitive statecraft are more or less similar.

The terms "irregular warfare" and "competitive statecraft" both have merit and utility in our current discourse. While the term "political warfare" had a good run in the Cold War, it has become problematic for two reasons. First, "political warfare" implies the exclusive use of political tools designed to undermine the governments of adversaries. However, for present and future contexts, "political warfare" would need to include the full spectrum of activities like hardening infrastructure and economic coercion. Second, the term "political warfare" has been repeatedly abused and adopted by US politicians, who have expressed frustration over—and contributed to—the polarization and tactics between political parties.[44]

Though the term "competitive statecraft" may dilute the call for action implied in the word "warfare," it may be more palatable in the current environment. Regardless, if the US continues to underprioritize nonmilitary means and treat them with anything less than the urgency implied in warfare, it will fail in its efforts to prevail in the current strategic competition.

Traditionally, there are four primary instruments of power: "information (words and propaganda), diplomacy (negotiations and deals), military (weapons, training, and violence) and economics (goods and money)."[45] Each instrument has associated tools that provide either positive or compelling incentives; or conversely, they can be used to thwart, punish, or deter. This book devotes chapters to these instruments to explain the tools available for states to craft their respective strategies.

In the information domain, key tools include information manipulation, infrastructure destruction, and sharing or declassifying foreign intelligence.[46] Agreements of all kinds (international conventions, treaties, formal alliances, memoranda of agreements, and understanding), as well as the use of political platforms to promote a particular narrative, are some of the core tools of diplomacy. Instead of devoting a separate chapter to diplomacy, we incorporate agreements into discussions of information, military, and economic statecraft. Public diplomacy and the promotion of particular narratives fall under information statecraft. Military statecraft (beyond the use of force) comprises activities such as multinational exercises, training and equipping foreign forces, professional military education, and a variety of security cooperation programming, including foreign military sales and humanitarian assistance.[47] In the economic domain, the tools are those related to trade and capital—both of which have positive and negative options—as well as domestic policies that facilitate their use.[48]

We propose a new instrument of power: resilience. Resilience enables a state to withstand IW approaches employed against it while absorbing and recovering from other shocks. Given the twin threats outlined in the 2022 NSS—the traumatic effects of climate change and strategic competition—the elevation of resilience to a distinct instrument of power is warranted.

Power, Influence, and Legitimacy

Clearly, there are many IW activities for competitors to pursue, but to what end? We argue that today's competition is ultimately about achieving relative power, influence, and legitimacy in the international system. These are necessary to shape and lead an international order that defends the interests and promotes the values of the most capable major power(s).

Figure 3. Definitions of IW Objectives.

TERM	DESCRIPTION
Power	Ability to coerce
Influence	Ability to convince
Legitimacy	Domestic and global population perceives that US actions and institutions are more right and appropriate than those of competitors

Power has long been central to international relations. It is a relational and causal concept,[49] and "defined to include all relationships in which someone gets someone else to do something that he or she would not otherwise do."[50] Power is a function of the "distribution of capabilities between states."[51]

Realists, from Kenneth N. Waltz to John J. Mearsheimer, believe the balance of power in the international system is a key determinant of its stability. Although capabilities are usually tangible factors like military forces and economic resources, the more recent literature on soft power emphasizes the importance of non-material leverage, like diplomatic relationships and cultural sway.[52] This book aligns with Ruth Zimmerling's definition of power: "the ability to get desired outcomes by making others do what one wants."[53] IW campaigns achieve and affect such power through measures usually below the threshold of armed conflict.

Whereas there are tomes written about power, influence is far less researched and often poorly defined.[54] Scholars also differ in how they perceive the relationship between power and influence.[55] Some view these terms as synonymous, defining both as the ability of one actor to get another actor to do something that they would not otherwise do.[56] Alternatively, others believe that influence involves a change that one actor imposes on another, but power is distinct in that it is the ability to exert that influence.[57] Thus, "power is something which can be possessed while influence is something that can be exerted."[58] But we subscribe to a third school of thought[59] that defines power as the overt ability to affect others' behavior, and influence as

> the "ability" to affect others' beliefs, that is, their knowledge or opinions either about what is or about what ought to be the case, about what is (empirically) true or false or what is (normatively) right or wrong, good or bad, desirable or undesirable.[60]

According to Mark C. Suchman, legitimacy is "a general perception or assumption that the actions of an entity are desirable, proper, or appropriate within some socially constructed system of norms, values, beliefs, and definitions."[61] Underlying much of the legitimacy literature is the idea that "a state is more legitimate the more that it is treated by its citizens as rightfully holding and exercising political power"[62] Legitimate states are peaceful and stable, while illegitimate states are the cause of violence and war.[63] Socially constructed norms shape the degree to which actions are deemed legitimate or not.[64] So, when judging the legitimacy of state actions, ranging from the use of force to the commitment to mitigate climate change, that reasoning is framed by international norms and expectations about the rights and responsibilities of great powers.[65]

Thus, power can be most simply defined as the ability to coerce, influence as the ability to convince, and legitimacy as the collective belief among a relevant population that something is rightful. There is a symbiotic and reinforcing relationship between power, influence, and legitimacy. The deliberate and measured use of power and influence can shape what a relevant population sees as legitimate. But how does a country obtain and sustain legitimacy, power, and influence? In strategic competition, the answer lies either in bolstering one's own legitimacy, power, and influence or in diminishing those of a competitor.

Book Overview

This book is deliberately broad in its scope. It is meant to help students and practitioners in the national security community to reframe their thinking about strategic competition and America's place in global politics. Past American strategies have neglected the role that information, resilience, and economics play in generating power, influence, and legitimacy. Why has the US approached strategy in this way? Chapter 2 explores this question, attributing it in large part to American strategic culture. Why is the current US approach a problem? Chapter 3 explains why US competitors do things differently by examining their respective strategic culture. It also examines the effects of culture on how they harness myriad tools to their advantage. Can the US change its approach despite its strategic culture? Yes, it has done so before. US strategy in the Cold War, as explained in chapter 4, suggests it is possible. How did the US change its strategic approach despite its strategic culture?

So, what should the US prioritize to address the challenges of strategic competition in an era of crises? Chapter 5 highlights the powerful role of military statecraft. Instead of preferencing force and prioritizing preparation for the most dangerous enemy course of action, the non-kinetic uses of the military should be central to US grand strategy. Chapter 6 explains the promise and peril of economic statecraft—often an alternative to the use of military force. How can the US use economic statecraft to leverage its position? The internet age has cemented information as foundational to IW; chapter 7 explains the dangers posed by mis- and disinformation and its harmful impact on democracies. What are potential actions the US can take to address these threats? What needs to be a key component of any future grand strategy? Chapter 8 argues that resilient states can better withstand the crises that arise in an era of strategic competition; hence, resilience should be at the core of any future grand strategy.

With long-term assessment in mind, how does one measure progress toward gains (or losses) in power, influence, and legitimacy? Chapter 9 examines the literature that addresses this question, suggesting a mosaic approach to measuring progress in the four domains (military, economic, information, and resilience). The final chapter offers concluding thoughts and recommendations for the way ahead.

Introduction 17

Notes

1. White House, *National Security Strategy*, (October 2022), 6.
2. Martel, "America's Dangerous Drift"; and Talent and Hegseth, "America's Strategic Drift."
3. White House, *National Security Strategy*, 2.
4. Kissinger, *On China*; Luttwak, *The Rise of China*; Allison, *Destined for War*; Yoshihara and Holmes, *Red Star over the Pacific*; Li, *The Rise of China, Inc.*; Moore, *China's Next Act*; Shirk, *Overreach*; and Dikötter, *China After Mao*.
5. White House, *2015 National Security Strategy*, ; White House, *2017 National Security Strategy*; US Department of Defense, *Summary of the 2018 National Defense Strategy of the United States of America*, 1,; White House, [2022 *National Security Strategy*.
6. Giske, "Resilience in the Age of Crises."
7. Watson and Tucker, "These Are the US Military Bases."
8. Health Canada, "Public Health Update."
9. Paul et al., *The Role of Information in US Concepts for Strategic Competition*, 2.
10. Ashford and Cooper, "Assumption Testing."
11. Bensahel, "Darker Shades of Gray"; Atlantic Council, "Today's Wars"; and Derleth, "Great Power Competition."
12. US Joint Chiefs of Staff, *Joint Concept for Competing*.
13. White House, [2022] *National Security Strategy*, 11.
14. Clausewitz, *On War*, 88.
15. When considering the elements of power holistically, this refers to Hans Morgenthau's elements of power which include population, territory, natural resources, industrial capacity, agricultural capacity, military strength, bureaucratic efficiency, leadership, type of government, social cohesiveness, reputation, foreign support, and diplomacy. See Morgenthau, *Politics Among Nations*, 106.
16. Baldwin, *Economic Statecraft*; Jones, "The Future of Competition"; and Ucko and Marks, *Crafting Strategy for Irregular Warfare*, 10.
17. Jones, "The Future of Competition."
18. Jones.
19. Jones, *Three Dangerous Men*, 5.
20. Lippman et al., "Biden's Era of 'Strategic Competition.'"

21. White House, *National Security Strategy*, 8.
22. Overfield, "Biden's 'Strategic Competition' Is a Step Back."
23. Wasser and Pettyjohn, "Why the Pentagon Should Abandon."
24. Nexon, "Against Great Power Competition."
25. US Joint Chiefs of Staff, *Joint Concept for Competing*, 1.
26. Mittelmark, "Playing Chess with the Dragon."
27. Mazarr, *Mastering the Gray Zone*, 58; Atlantic Council, "Adding Color."
28. Krishnan, "Fifth Generation Warfare," 28.
29. Gray, *Categorical Confusion?*
30. Arquilla, *Insurgents, Raiders, and Bandits*.
31. Marks and Ucko, "Counterinsurgency as Fad."
32. US Naval War College, "Center on Irregular Warfare & Armed Groups."
33. US Department of Defense, *Summary of the Irregular Warfare Annex*.
34. Ucko and Marks, "Redefining Irregular Warfare."
35. Krieg and Rickli, *Surrogate Warfare*.
36. Jones, "The Future of Competition."
37. US Department of Defense, *DOD Dictionary 2008 (Through 2021 edition)*
38. Tucker, "As Irregular Warfare Comes."
39. Jones, *Three Dangerous Men*, 5.
40. Webster, "#Reviewing Three Dangerous Men."
41. Kennan, "Foreign Relations of the United States, 1945-1950."
42. Bilms, "What's in a Name?"
43. Shaw, "In Defense of Competition."
44. Parker and Eisler, "Political violence."
45. Lasswell, *Politics*; and Steil and Litan, *Financial Statecraft*, 1.
46. Le Gargasson and Black, "Reflecting on One Year of War."
47. Defense Security Cooperation Agency, "Security Cooperation Overview."
48. Baldwin, *Economic Statecraft*.
49. Dahl and Stinebrickner, *Modern Political Analysis*, 12.
50. Baldwin, *Economic Statecraft*, 18.
51. Waltz, *Man, the State, and War*, 101.
52. Nye, "Soft Power"; Nye, "Public Diplomacy"; and Yun, "An Overdue Critical Look."
53. Zimmerling, *Influence and Power*, 141.
54. Zimmerling, *Influence and Power*; and Meierding and Sigman, "Understanding the Mechanisms."
55. Moyer et al., "Power and Influence," 6–7

Introduction 19

56. Dahl and Stinebrickner, *Modern Political Analysis*; and Baldwin, *Economic Statecraft*.
57. Morgenthau, *Politics among Nations*; Waltz, *Theory of International Politics*; and Cartwright, "Influence, Leadership, Control."
58. Moyer et al., "Power and Influence," 6.
59. Moyer et al., 6–7; Zimmerling, *Influence and Power*; and Wrong, *Power*.
60. Zimmerling, *Influence and Power*.
61. Suchman, "Managing Legitimacy," 574.
62. Gilley, "The Meaning and Measure of State Legitimacy."
63. Woodward, "Can We Measure Legitimacy?"
64. Reus-Smit, "Power, Legitimacy, and Order."
65. Reus-Smit.

Chapter 2

American Strategic Culture

> Know the enemy; know yourself, and you will meet with no danger in a hundred battles. If you do not know the enemy, but you know yourself, then you will win and lose by turns. If you know neither the enemy nor yourself, you will lose every battle certainly.[1]
>
> Sun Tzu

Introduction

The idea that certain groups, whether states, military organizations, or ethnicities, have a preferred method of waging war is not revolutionary. Historians and strategists have observed these differences at least since the Peloponnesian War.[2] Jack L. Snyder first introduced the idea of strategic culture and its role as a critical variable in political and military decision-making in "The Soviet Culture: Implications for Limited Nuclear Operations." In it, he describes Soviet strategic thinking as the product of a "unique strategic culture" encompassing multiple factors, most notably history and geography.[3] Since then, strategic culture, its inputs, utility, and even its existence, have generated much research and debate. This chapter extends this conversation with particular regard to the impacts of American strategic culture in an era of crises and strategic competition.

The predominant American strategic culture does not have a predisposition for long-term geopolitical competition. American history and geography have combined to create a strong cultural preference for clear dichotomies: black and white, good and evil, and war and peace. However, these do not represent neither the current international environment in which the competition is unfolding nor the ongoing reality of long-term transnational threats. Given the reticence to embrace this reality, there is little likelihood of a positive outcome for the US without a concerted effort to change.

The chapter begins by examining the evolution of strategic culture scholarship, its relative utility, and the challenges that arise when competing cultures exist simultaneously. It also considers circumstances that create cultural change and then examines several of American strategic culture's most commonly attributed characteristics. The chapter concludes with an analysis of the implications of these traits.

The Half-century Evolution of Strategic Culture Scholarship

This section first highlights the challenges of using strategic culture for theory building and theory testing. It then provides an overview of the hierarchy of strategic culture explaining the relevant subcultures, both political and military. Finally, it addresses the factors that inhibit changes to strategic culture.

Methodological Considerations

Snyder's article proposes that Soviet decision-making concerning nuclear weapons differs substantially from the American calculus due to profound cultural differences. He argues:

> Soviet and American doctrines have developed in different organizational, historical, and political contexts, and in response to different situational and technological constraints. As a result, the

Soviets and Americans have asked somewhat different questions about the use of nuclear weapons and have developed answers that differ in significant respects.[4]

This analysis diverges radically from the "rational actor model" that often underpins American nuclear policy decision-making. It also pries open the tightly shut black box associated with realist theories of international relations.[5] In so doing, strategic culture adherents spawned a prolific and, at times, less than collegial debate on the concept that has now spanned almost half a century.[6]

Perhaps the most consequential progenitor of the idea is Colin S. Gray, whose definition of strategic culture—"the persisting socially transmitted ideas, attitudes, traditions, habits of mind, and preferred methods that are more or less specific to a particular geographically based security community that has had a unique historical experience"—continues to be used today.[7] This work adopts Gray's definition. However, other proposed definitions would suffice, given their agreement on history, geography, and political systems as the most salient variables.

Gray's views on strategic culture still resonate with practitioners because they are pragmatic. Explaining his views on strategic culture's role in policymaking, Gray observes, "strategic culture can be conceived as a context out there that surrounds, and gives meaning to, strategic behaviour [sic], as the total warp and woof of matters strategic that are thoroughly woven together, or as both."[8] Further, he argues all strategy and policymaking are inherently "encultured," meaning there is no aspect of decision making or strategic formulation free of cultural context or influence.[9] Put another way, culture is as ambient a condition for human interaction as water is to the lives of fish. There is no activity it does not encompass.

Acknowledging the concept of enculturation renders arguments about strategic culture explanations for state behavior versus those of realists or constructivists moot. Given that all human behavior is encultured, there can be no international relations theory that is culturally neutral.

However, it is impossible to disentangle the disparate threads that comprise strategic culture and determine which will drive a particular choice in a specific circumstance, confounding theorists and practitioners wishing to adopt a theoretical framework that includes strategic culture as a variable in decision-making.

Given these substantial obstacles, the concept of strategic culture has not been universally embraced.[10] Critics rightly point out that the relationship between culture and behavior is not linear. Acknowledging that a state's behavior differs from that of another because of geographic, historical, and political circumstances is one thing, but using those differences to predict strategic choices has proven elusive.

Alistair Iain Johnston summarized the critiques of early strategic culture scholarship, stating:

> [T]he concept of strategic culture was extremely unwieldy. Technology, geography, organizational culture and traditions, historical strategic practices, political culture, national character, political ideology, and even international systems structure were all deemed relevant inputs into this amorphous strategic culture. Yet, arguably, these variables are different classes of inputs; each could stand by itself as a separate explanation of strategic choice. If 'strategic culture' is said to be the product of nearly all relevant explanatory variables, then there is little conceptual space for a non-strategic culture explanation of strategic choice.[11]

His critique that strategic culture explanations are so broad as to be non-falsifiable and, therefore, useless in determining causality is compelling and should serve as a warning of the concept's limited applicability in practice. However, Gray's response to this critique is equally persuasive, arguing that strategic culture should remain a critical variable in strategic decision-making. In a direct response to Johnson's critique of strategic culture's utility, Gray argues that Johnston misses the point entirely or that perhaps Johnston's point is only an issue for academics and those concerned with building a theory of strategic culture. Gray bluntly asked,

"So what?" in response to Johnston's concerns.[12] The methodological difficulties associated with strategic culture do not negate its actual impact on strategic decision-making.

For practitioners, the non-falsifiability of strategic culture explanations is just one more thing that makes strategy so difficult. However, ignoring strategic culture's existence only creates blind spots. Therefore, decision-making must account for strategic culture, however amorphous that accounting is in practice.

A Hierarchy of Cultures
As problematic as enculturation is, it is not the only obstacle to incorporating the idea into strategic decision-making. Determining which culture is driving decision-making at a particular moment is equally challenging. One must acknowledge the existence of a hierarchy of multiple cultures operating simultaneously and sometimes in competition.

First, one must understand that strategic culture derives from the broader concept of political culture, which was introduced in the 1960s by Gabriel A. Almond and Sidney Verba,[13] who defined political culture as "composed of the attitudes, beliefs, emotions, and values of society that relate to the political system and political issues. These attitudes may not be consciously held but may be implicit in an individual or group relationship with the political system."[14] Strategic culture is a subculture of political culture.

Snyder defines "subculture" as "a subsection of the broader strategic community with distinct beliefs and attitudes on strategic issues."[15] Thus, within any cultural construct, one or more subcultures can exist and function independently. They can also interact and create effects on strategic decision-making that defy prediction. Much like the interplay of the variables in Clausewitz's trinity, the relationship among the extant subcultures is complex and non-replicable.

American political culture can be subdivided into subcultures or strands of thought regarding foreign policy. Walter Russell Mead argues that four prominent strains of political culture have asserted themselves periodically throughout American history.[16] They are:

- **Hamiltonian**: characterized by a strong national government and strong alliances;
- **Wilsonian**: characterized by a moral obligation and national interest to spread democracy;
- **Jeffersonian**: mainly concerned with safeguarding democracy at home;
- **Jacksonian**: mainly concerned with physical security and economic prosperity.[17]

Given this, as one attempts to determine the character of American strategic culture, one must also consider which of these strains of political culture held sway at the time.

Militaries have their own subculture. Jeffrey Legro defines military subculture as "the pattern of assumptions, ideas, and beliefs that govern how a military bureaucracy should wage war will shape state preferences and actions on the use of that means. Each military creates its own culture that sets priorities and apportions resources."[18]

Evidently, strategic culture encompasses multiple complementary and competing inputs. Thomas G. Mahnken observes, "One of the central challenges facing the scholar of any state's strategic culture lies in determining which institutions serve as the keeper and transmitter of strategic culture. Is it the state? The military as a whole? Or some subset of the military?"[19] He conceptualizes strategic culture as having three distinct yet interrelated levels. They are:

> [T]hose of the nation, the military, and the military service. At the national level, strategic culture reflects a society's values regarding the use of force. At the military level, strategic culture (or a nation's "way of war") is an expression of how the nation's military wants to

fight wars.... Finally, strategic culture at the service level represents the organizational culture of the particular service–those values, missions, and technologies that the institution holds dear.[20]

Mahnken's three-level formulation provides a valuable framework for the hierarchy and complexity of strategic culture. That said, one cannot help but wonder if this formulation constitutes a necessary but insufficient condition. It is reasonable to assume military culture will play an outsized role in determining when and how the US uses force to pursue its strategic goals, given the size of the DOD's budget relative to other departments. However, there is also the role of American diplomacy and US economic power to consider as elements of strategic culture.

For the purposes of this chapter, military culture will be treated as *primus inter pares* among the American strategic subcultures, but it will not be considered in isolation. To incorporate diplomatic and economic preferences into the discussion of American strategic culture, we will include Mead's four aforementioned strains of political culture.

Does Strategic Culture Change?
This leaves one further issue for consideration: does strategic culture change? And if so, how often? What causes these changes? Although strategic culture changes slowly, it does evolve. Innovations in technology, shifts in the international system, changes in threat perception, and significant shocks to the political system—such as major wars and economic catastrophes—drive this evolution.[21] Even in the case of an abrupt transformation of the governmental structure, strategic culture tends to endure. As chapter 3 illustrates, Russian strategic culture has maintained startling continuity over the past century despite the strategic shocks caused by wars, revolutions, and profound ideological shifts.[22]

There are several reasons for the persistence of strategic culture. First, it is based on multiple variables, most prominently history and geography. Although history is subject to interpretation, it provides a relatively constant high-level input to strategic culture.[23] Geography is immune to societal change and political revolutions. Thus, history and geography provide a reasonably stable cornerstone for a nation's strategic culture.

Second, and equally important, strategic culture exists within the realm of cognition. It is a belief system or schema that provides frameworks for understanding and decision-making.[24] As such, it is highly resistant to change. As Thomas S. Kuhn explains, humans will distort facts to ensure their preferred explanation fits within existing, agreed-upon paradigms.[25] Reluctance to acknowledge or assimilate discordant information makes strategic culture—like political culture and military service culture—slow to change, except in the wake of significant external shocks that cannot be ignored or shoehorned into an existing cognitive framework.

The US experience in World War II exemplifies this type of shock. At the outset, the US was content to remain largely disconnected from European power politics. After the war, however, America stepped into a position of global leadership and, alongside its allies, created a new world order that it has maintained for decades. Gray argues that this experience precipitated a dramatic change in US strategic culture as Americans "endorsed the idea that the United States should be, or had to be, a permanent guardian of the international order."[26]

Thus, significant shocks have the potential to move the US from one strain of Mead's four categories of preferred foreign policy to another. The American response to the Japanese attack on Pearl Harbor and the 9/11 attacks have evinced this phenomenon. After the Pearl Harbor attack, many Americans shifted from a belief in isolationism—which Mead describes as a Jeffersonian preference mainly concerned with safeguarding democracy at home—to a Hamiltonian one, which is predicated on a strong national government and a robust system of alliances.[27] Similarly, the 9/11 attacks shifted the US into a Wilsonian strain of foreign policymaking, as the Bush administration became principally concerned with exporting democracy, taking unilateral action to do so.[28]

Michael J. Boyle and Anthony F. Lang argue that the US is uniquely prone to particularly strong reactions to strategic shock, resulting in near-instantaneous and predictable shifts in American strategic culture. They argue that "the instinct to remake societies in its own image is latent but inherent within American strategic culture and activated by strategic surprises that present a new type of ideological threat."[29] They further assert that this reflexive reaction to strategic surprise is "broadly vindicationist in the sense that it seeks to validate American values and institutions, and spread them by force if necessary."[30] Thus, Boyle and Lang argue that a strategic shock to the US will elicit a shift to a Wilsonian paradigm in American strategic culture.

The American response to strategic shocks is an extreme example of the impetus for change in strategic culture. Most cultural shifts occur slowly over time and in response to large quantities of new information that do not comport with the current paradigm. Sudden conversions in strategic culture are far rarer.[31]

AMERICAN STRATEGIC CULTURE

Having reviewed the definition, composition, and debates surrounding strategic culture, we now turn to a discussion of the prominent characteristics of American strategic culture and their probable impacts on the US approach to strategic competition. The preceding sections have shown the difficulty, if not the impossibility, of employing strategic culture as a predictive tool. Nonetheless, it can be a useful method for bounding a state's calculation of acceptable choices. Gray argues,

> It is not that culture is the driving force behind a decision. There are many inputs. That said, whatever the mix of factors in the decision—all of the people and organizations within which they function are more or less distinctly encultured.[32]

By understanding the contours of a state's culture, one can develop a sense of the boundaries of acceptable choices. Jeannie L. Johnson contends, "Cultural influences do not provide a clear-cut script for action, but tend to bound our beliefs about the range of effective and appropriate options available."[33] With these caveats in mind, after defining the characteristics of American strategic culture, we will address their implications, particularly as they relate to IW.

Introduction

American strategic culture is a function of its history, and geography. The unique circumstances of its founding and geographic isolation from its European progenitors produced a strategic culture that is fundamentally different from that of its closest allies, a fact that continues to create discord and misunderstandings two hundred and fifty years later. American strategic culture rests on an immutable faith in American exceptionalism and the moral rectitude of democracy.[34] The US also differs from the European realpolitik tradition in its belief that peace is a natural condition and war is an aberration.[35] As such, the US manages war as a crusade against evil, intending to restore the natural order of things.[36] Finally, American strategic culture encompasses a fundamental optimism cultivated from the historical narrative of conquering the frontier; for Americans, all problems have solutions if one applies the correct technology and works hard enough.[37] Given time and adequate resources, progress will ultimately prevail. This belief has produced a bias for action over analysis and reflection, which penetrates deeply into American strategic, military, and subordinate (armed services) cultures.[38]

The following section explores each of these threads and their position in the warp and weft of American strategic culture, as well as their implications for the US today. Sun Tzu suggests self-knowledge is required to win the battle. In the American effort to win without fighting, a dose of self-knowledge may not guarantee success, but it will help forestall failure.

Impacts of Geographic Isolation

There is a common saying in real estate that three things matter: location, location, and location. This is also true of American strategic culture. Otto von Bismarck once quipped, "The Americans are very lucky people; they are bordered to the North and South by weak neighbors and to the East and West by fish."[39] As a result, the US developed its national identity without ever facing an existential crisis from abroad.

In George Washington's farewell address, he admonished the young country to "avoid entangling alliances."[40] The fact that doing so was a viable choice made it unique among nations and significantly impacted the development of American strategic culture. Freedom from outside interference allowed the US to develop a strain of isolationist foreign policy preference that periodically reasserts itself (Jeffersonianism).[41] As Gray observes, "The United States is an insular power of continental size."[42] America not only wanted to avoid European realpolitik, it actually also could. This freed it to fulfill its "manifest destiny," the powerful nineteenth-century narrative that it was the duty of the US to expand westward and conquer the frontier.

The American experience of the seventeenth, eighteenth, and nineteenth centuries—winning independence against the world's most powerful empire and mastering the vast, varied, and inhospitable terrain of the American continent—imbued the American psyche with the belief that anything is possible. As a result of its pioneering history and problem-solving approach, Americans undervalue the study of history and philosophy in favor of "practical knowledge."[43] Americans are cowboys and engineers, not philosophers, creating an action bias in the fabric of American strategic culture. "The military puts this impulse on steroids."[44]

Practically speaking, Americans are impatient and incautious, and the military, despite the misgivings of its most senior general officers, can be even more so. Johnson opines on this action bias within the US military and government, pointing out that it is coupled with an "obsession" with metrics and quantification.[45] However, not everything can be quantified; culture, diplomacy, and foreign policy defy measurement on a Likert scale (a scale commonly used in questionnaires to gauge attitudes, values, or opinions). This produces blind spots in analysis and increases the risk of failure.

Another effect of American isolation is a propensity for mirror imaging. Although its Declaration of Independence famously reads that "all men are created equal,"[46] equal does not mean identical. Nor does it imply that everyone values the same things. This nuance often escapes US policymakers, who tend to mirror-image the desires of others. Gray observes,

> in part courtesy of the a-, or even anti-, historical training of American policymakers, American national security policy tends typically to be dominated by people who are truly expert only in inappropriate domestic political matters.... Harvard Law School, Wall Street, or a governor's state house do not prepare one well, in general, for coping with the surviving graduates of Stalin's "Great Purge" of the 1930s.[47]

Although the reference to Joseph Stalin's purges is clearly dated, the underlying sentiment remains valid. Further, geographic isolation reduces the level of American cross-cultural fluency, particularly regarding language:

> But since most Americans only speak one language, they are usually dependent on finding English speakers or translators. Once they have succeeded in their search, Americans are likely to believe that the problem of language is solved.[48]

The idea that an interpreter will successfully navigate all nuances of understanding created by cultural differences is, at best, naïve and, at worst, dangerously misguided. Nonetheless, it is a facet of American strategic culture.

Finally, geographic isolation allowed the US to develop an "astrategic" strategic culture.[49] Gray argues that America has had little need to develop genuine strategic competence due to its geographically secure location and tremendous resource endowment. Instead of developing strategy, Americans prefer to engage in military planning, a technical exercise in resource allocation, not in strategy.[50] More worrisome is that most US military officers regard these activities as interchangeable.[51]

In the minds of many Americans, isolation facilitated hegemony and, along with it, the creation of the world's largest and most technologically advanced military. However, geographic isolation no longer protects us as it did a century ago. To win without fighting in an era of crises and competition, the US must acknowledge its cultural blind spots and their strategic implications.

The City on the Hill: American Exceptionalism
Americans see their country as exceptional. The term "American exceptionalism" is generally attributed to Alexis de Tocqueville, who argues that the US perceives itself as fundamentally different because of its unique origins, national credo, political institutions, and religious heritage.[52] A component of this exceptionalism is a strain of seventeenth-century Puritan fundamentalism that still resonates. In a 1630 sermon, John Winthrop famously declared, "For we must consider that we shall be as a city on a hill. The eyes of all people are upon us...."[53] This imagery manifests in several ways; the US holds a nearly evangelical faith in the superiority of democracy over all other forms of government, makes (negative) moral judgments about other political systems, and asserts that liberal values should underpin the global order.[54]

Benjamin Brice explores the role of American exceptionalism in the 2003 invasion of Iraq. He asserts that American leaders explicitly used liberal ideals to underpin their foreign policies and argues that although there was a mixture of motives involved in the invasion, all were undergirded by "moral considerations."[55] Brice concludes that although Americans do not believe these ideals are exclusive to the US, they believe their country does have "a unique responsibility to promote them."[56]

Mead refers to this sense of responsibility to promote democracy abroad as Wilsonianism.[57] Boyle and Lang argue that Wilsonianism constitutes a latent instinct to remake societies in the American image—

one that activates in the face of strategic shocks and ideological threats.[58] They conclude that this Wilsonian way of intervention has a dark side, "as policymakers often cast those who oppose their plans as malcontents opposed to the natural political order and punish them."[59]

Thus, American exceptionalism and its reflexive predilection to rely on its military can lead America into costly wars: with states whose cultures Americans do not understand, and whose desires policymakers largely deduce through mirror imaging, and whose interests they see only in relation to the US. H. R. McMaster refers to this phenomenon as "strategic narcissism."[60] By ignoring the mandate for cultural analysis, the US diminishes its power, influence, and legitimacy. Strategic competition requires cultural understanding as much, if not more, than conventional military capabilities.

War as Evil and Aberrant
Another facet of rejecting European power politics is the American belief that war is evil unless fought for pure purposes and that peace is the natural state of humanity.[61] In *The Soldier and the State*, Samuel P. Huntington observes:

> The American tends to be an extremist on the subject of war: he either embraces war wholeheartedly or rejects it completely....war must be either condemned as incompatible with liberal goals or justified as an ideological movement in support of those goals.[62]

Illustrating Huntington's hypothesis, President Woodrow Wilson entered World War I on a quest to bring democracy and Christian values to politics in a way that would "perfect humanity" and end war.[63] At its conclusion, he attempted to expand American liberal ideology by establishing the League of Nations. World War II was similarly cast as an "epic struggle of democratic heroes stubbornly determined to slay the Nazi monster."[64] At its conclusion, the Allies founded the United Nations (UN) and remade the "monsters" as democracies. More recently, the US invasion of Iraq was undertaken to oust Saddam Hussein, a member of President George

W. Bush's "Axis of Evil."[65] The resulting regime change intended to create a functioning democratic state.

War and Peace as Distinct
Further complicating American prospects for success in strategic competition is the bifurcation of war and peace as entirely distinct concepts. This makes it difficult to engage successfully in the gray-zone activities required to prevail in strategic competition. Writing at the end of World War II, Kennan observed,

> We have been handicapped, however, by a popular attachment to the concept of a basic difference between peace and war, by a tendency to view war as a sort of sporting context outside of all political context, by a national tendency to seek for a political cure-all, and by a reluctance to recognize the realities of international relations—the perpetual rhythm of [struggle, in and out of war.][66]

Seventy-five years later, American strategic culture remains deficient. In fact, the problem has worsened in the intervening decades. The DOD has increased its budget and capacity to employ lethal effects using cutting-edge technology, whereas the Department of State (DOS) has taken substantial cuts to its comparatively paltry funding.[67] The space between war and peace, where much of the maneuvering of strategic competition occurs, is not well provisioned due to these enduring strategic culture preferences.

The next chapter explores the strategic cultures of America's current adversaries and their implications,[68] offering valuable lessons, mainly because they differ so dramatically from that of the US. In particular, the line Americans draw between war and peace is fuzzy or even nonexistent to competitors. Sun Tzu's concept of winning without fighting is often touted as the antithesis of American strategic thinking. That said, Sun Tzu's ideas are not even the most radically divergent from those of American strategic culture. For that, we have to turn to Kautilya.

Far less known than Sun Tzu in the West, Kautilya came to preeminence in what is now India following the death of Alexander the Great. His seminal work, *The Arthashastra*, aimed to guide the Mauryan emperors.[69] Whereas Clausewitz famously described war as a continuation of politics by other means, Kautilya proclaims diplomacy to be a subtle act of war in which the actual purpose of negotiation is to weaken one's enemy.[70] He follows this with his unique doctrine of "silent war"—a concept completely antithetical to the American way of war, explaining:

> Silent war is a kind of warfare with another kingdom in which the king and his ministers—and unknowingly, the people—all act publicly as if they were at peace with the opposing kingdom, but all the while secret agents and spies are assassinating important leaders in the other kingdom, creating divisions among key ministers and classes, and spreading propaganda and disinformation.[71]

Kautilya's work is far more than an obscure ancient text. It underpins the contemporary strategic culture of the Indian subcontinent. In 2020, Michael Liebig observed, "Kautilya is in the DNA of India's security"[72]; and his version of realism is core to Indian strategic culture.[73] Though assassinations and disinformation can often be counterproductive in the quest for legitimacy, American strategists must recognize the fluid relationship between war and peace.

An Enduring Preference for Overwhelming Force

More than simply framing conflict as a struggle between good and evil, US military culture and its preference for using overwhelming force has also impacted the American propensity to engage (or not engage) in foreign wars.[74] From Sherman's March to the Sea in the American Civil War to the Weinberger-Powell Doctrine and its requirement for overwhelming force, American military culture prefers large-scale conventional combat operations to IW.[75] With respect to this preference, Gray observes, "questions pertaining to the actual employment of force, and particularly of limited force, have been deemed secondary to the marshaling of muscle."[76]

IMPLICATIONS

From the preceding paragraphs, one can easily deduce that American strategic culture places the US at a disadvantage for multiple reasons. Many stem from the same geography and history, ironically, that allowed the US to rise to its current level of preeminence. Returning to Sun Tzu's admonition to know oneself, US policymakers should begin by examining their strategic cultures and attempting to map the "water in which they swim." This will facilitate an understanding of how culture bounds their choices. The underwhelming outcomes of America's recent wars should provide sufficient impetus for such self-examination. Though rigorous introspection will likely not induce wholesale change to American strategic culture, it may illuminate areas of vulnerability where the reflexive need for action may be misplaced.

Strategic competition is not war. That said, the contours of the current geopolitical environment, with its conflicts of military, information, and economic varieties, showcase the weaknesses of America's dominant strategic and military cultures. Mark Galeotti opines, "wars without warfare, non-military conflicts fought with all kinds of other means, from subversion to sanctions, memes to murder, may become the new normal."[77] Although the dominant paradigms within American strategic and military culture are ill-suited to the current environment, there are options to mitigate these mismatches.

CONCLUSION

For the US, winning will be difficult. Culturally, America is primed to fight conventional wars against "evil empires" rather than engage in gray-zone activities to counter the actions of adversaries whose governmental structure and worldviews differ radically from their own. That said, there are vibrant subcultures within the military and government whose mindset and competencies suit the contours of strategic competition. Such US government personnel can be found in the State Department's Foreign Service, US Special Operations Forces, and the operational arms of the US intelligence communities are just a few of them. To meet the current challenge, we need to reinvigorate these communities.

Notes

1. Sun Tzu, *The Art of War*, 56.
2. Gilchrist, "Why Thucydides Still Matters," 1.
3. Snyder, *The Soviet Strategic Culture*, 5.
4. Snyder, v.
5. Borrie, "Human Rationality and Nuclear Deterrence," 10.
6. For a relatively recent installment in the debate over strategic culture, see the exchange between Frank Hoffman and Antulio Echevarria in Hoffman, "Review: Strategic Culture And Ways Of War."
7. Gray, *Modern Strategy*, 10.
8. Gray, "Strategic Culture as Context," 51.
9. Gray, 56.
10. Johnston, "Thinking about Strategic Culture," 36.
11. Johnston, 37.
12. Gray, "Out of the Wilderness," 9.
13. Almond and Verba, *The Civic Culture*, 11–14.
14. Kavanaugh, *Political Culture*, 53.
15. Snyder, *The Soviet Strategic Culture*, 10, 12.
16. Mead, *Special Providence*, xvii.
17. Mead.
18. Legro, "Military Culture and Inadvertent Escalation in World War II," 117.
19. Mahnken, "United States Strategic Culture," 4.
20. Mahnken, 5.
21. Dueck, *Reluctant Crusaders*, 34.
22. Eitelhuber, "The Russian Bear," 4.
23. Dueck, *Reluctant Crusaders*, 37.
24. Larson, "The Role of Belief Systems and Schemas in Foreign Policy Decision-Making," 29.
25. Kuhn, *The Structure of Scientific Revolutions*, 45.
26. Gray, "National Style in Strategy," 22.
27. Mead, *Special Providence*, vi.
28. Mead, 5.
29. Boyle and Lang, "Remaking the World in America's Image," 2.
30. Boyle and Lang.

31. Larson, "The Role of Belief Systems and Schemas in Foreign Policy Decision-Making," 29.
32. Gray, "Out of the Wilderness," 8.
33. Johnson, "Fit for Future Conflict?" 187.
34. Owens, "American Strategic Culture and Civil-Military Relations," 46.
35. Gray, "National Style in Strategy," 46.
36. Huntington, *The Soldier and the State*, 8.
37. Stewart and Bennett, *American Cultural Patterns*.
38. Johnson, "Fit for Future Conflict?" 189.
39. Reuter, "An Isolationist United States?"
40. George Washington's Mount Vernon, "Washington's Farewell Address, 1796."
41. Mead, *Special Providence*, xvii.
42. Gray, *The Geopolitics of Super Power*, 45.
43. Johnson, "Fit for Future Conflict?" 189.
44. Johnson, 193.
45. Johnson, 194.
46. The National Archives, "Declaration of Independence."
47. Gray, "Strategic Culture as Context," 46.
48. Stewart and Bennett, *American Cultural Patterns*, 45.
49. Gray, "National Style in Strategy," 25.
50. Gray.
51. The reference for this note is the authors' personal observations over close to 100 years in the US military collectively, the majority of which were spent in strategic planning assignments.
52. Tocqueville, *Democracy in America and Two Essays on America*, 36–37.
53. Heike, *The Myths That Made America*, 152.
54. Koh, "America's Jekyll and Hyde Exceptionalism," 112.
55. Brice, "A Very Proud Nation ," 63.
56. Brice , 61.
57. Mead, *Special Providence*, xvi.
58. Boyle and Lang, "Remaking the World in America's Image," 2.
59. Boyle and Lang .
60. McMaster, *Battlegrounds*, 15.
61. Gray, "National Style in Strategy," 43.
62. Huntington, *The Soldier and the State*, 151.
63. Knock, *To End All Wars*, 6.
64. Schrijvers, "War Against Evil," 2.
65. President George W. Bush, "State of the Union Address 2003," 109.

66. Kennan, "Policy Planning Staff Memorandum."
67. Katz, "Unfrozen."
68. White House, *2022, National Security Strategy*, 6.
69. Kumar, "The Arthaśāstra," 2.
70. Boesche, "Kautilya's Arthasastra," 20.
71. Boesche, 22.
72. Liebig, "India's Strategic Culture," 7.
73. Liebig, 4.
74. Ulrich, "Overwhelming Force," 4.
75. Record, "Back to the Powell-Weinberger Doctrine?" 81.
76. Gray, "National Style in Strategy," 26.
77. Galeotti, *The Weaponisation of Everything*, 9.

Chapter 3

Competitor Approaches to Irregular Warfare

> China and Russia conquer through irregular-war strategies. That works because they disguise war as peace, until it's too late.[1]
> Sean McFate

Introduction

As the last chapter explained, US strategic culture often inhibits policymakers from developing and implementing a coherent IW strategy. Authoritarian regimes, however, prioritize centralized and coordinated IW campaigns as integral to strategic competition. Despite perceiving America as the primary source of instability to both their regimes and their vision for a multipolar world, China and Russia seek to avoid direct confrontation; rather, they conceptualize a long-term struggle against an American adversary whose response is often short term and piecemeal.[2] Therefore, to better understand and ultimately defeat Russian and Chinese IW approaches, this chapter explores their use of military, economic, and information statecraft to erode American power, influence, and legitimacy while developing their national resilience capabilities to defend themselves from US influence.[3] Other strategic competitors, such as Iran and North Korea, similarly prioritize coordinated IW approaches in their own way and contexts. Although this chapter focuses exclusively on China and Russia's approaches, it does so to illustrate the two most serious IW threats to the international order and should not be construed as diminishing the IW potential of other competitors.

COMPETITOR STRATEGIC CULTURES AND IW

Whereas American strategic culture can impede commitment to IW, Chinese and Russian strategic cultures incorporate it. Although the unique histories and geographies of these competitors contribute in part to their approaches, the most critical variable is their regime type. Autocracies are more risk-acceptant of foreign policy adventures and more willing to use IW and violence.[4] Conversely, democratic regimes are more prepared to signal resolve on their most critical foreign policy issues and to fight and win major wars.[5] Thus, recent reforms that consolidate political power in the hands of Xi Jinping and Vladimir Putin suggest their regimes will become increasingly aggressive in their IW approaches. This section provides an overview of both Chinese and Russian strategic cultures.

Chinese Strategic Culture

China's perspective on internal and external security is underpinned by its strategic culture, which can be understood through various key pillars, starting with its history and geography. Historically, Chinese leaders conceived of their state as the Middle Kingdom; beyond mere geography, this term implied a role for Beijing at the cultural, political, and economic center of the world. For centuries, Chinese leaders defended themselves from neighboring enemies and foreign influence.

But this power and prestige began to wane upon the dawn of the First Opium War in 1839, until the end of the Chinese Civil War in 1949, which marks the Century of Humiliation. Domestically, China suffered from weak governance, disunity among elites, a rejection of technologies associated with Western culture—including those necessary for modern warfare—and a rampant drug trade.[6] It also experienced persistent defeats and occupations, which generated interwoven narratives of submission, invasion, interference, and overall humiliation later adopted by the Chinese Communist Party (CCP).[7] Today, these narratives precipitate hypersensitivity to foreign influence and threats to political unity,[8] lending support to Xi's centralization of authority and ambitious pursuit of the "China Dream" and "National Rejuvenation," CCP policy slogans for a return to great power status.[9]

A second pillar is traditional Chinese philosophy. Both Confucianism and Legalism are foundational to Chinese strategic culture. Confucianism emphasizes the importance of morality, hierarchy, and social harmony. Legalism profoundly influenced the establishment of the Chinese state, which demanded the centralization of power and strict legal control. Throughout its history, China has had a strong central bureaucracy (the *keju* system) but a weak civil society.[10] Chinese strategic culture and military institutions are also greatly influenced by Sun Tzu's *The Art of War* (as well as other military classics) and its emphasis on winning without fighting and strategic deception, core elements of IW campaigns.[11]

A third pillar is the Chinese political structure; since strategic culture is a subset of political culture, the CCP is a major determinant.[12] Drawing primarily from the Century of Humiliation—but also from recent periods of upheaval like Mao Zedong's Cultural Revolution and the 1989 Tiananmen Square protests and massacre—the CCP considers its hold on power to be vulnerable to domestic instability and revolutionary elements, resulting in a heavy-handed approach to internal dissent.[13] This vulnerability is a key driver of Chinese strategic competition with the US. Indeed, in the wake of the 1989 collapse of Eastern European communist regimes, the CCP—like the Kremlin—remains convinced that the US is actively fomenting domestic unrest to topple all communist governments worldwide.[14] This demands a vigilant response.

The fourth pillar is the Chinese economy; since the end of the Cold War, China has transformed itself from a poor and isolated country into the world's second largest economy, greatly enhancing its power, influence, and legitimacy.[15] China frames its economic growth as a strategy of peaceful development, focusing on internal economic development to strengthen its national power. Its newfound wealth affects its strategic culture, strategy, and approach to IW. Economic statecraft has become a major element of China's IW toolset while also providing the foundation for other means of national power.

The final pillar is sovereignty; due to the profound effect of the Century of Humiliation, modern Chinese strategic culture emphasizes national sovereignty. Whereas sovereignty has remained a shared principle of international relations since the Peace of Westphalia in 1648, the CCP view of sovereignty has always been extreme and has only grown more so since Xi's rise.[16] In particular, it supersedes concerns over human rights and provides the foundation for China's principle of non-interference in other countries' internal affairs. China rejects the ideas underpinning the Responsibility to Protect,[17] the idea that if a part of a population is threatened, foreign powers have a responsibility to intervene and protect them. That said, China hypocritically exploits this principle to intervene in protection of Chinese ethnic minorities in neighboring countries. When China perceives its sovereignty being threatened, it is increasingly likely to fight, especially in Asia.[18]

Chinese and American strategic cultures are very different. Unlike the US, Chinese strategic culture conceives of a fluid relationship between war and peace; indeed, all statecraft is essential to achieving power as the primary objective in their never-ending "struggle." [19] Chinese strategic culture also lays the foundation for whole-of-nation efforts to build power, influence, and legitimacy—thereby redressing its core vulnerabilities, both actual and perceived, in an attempt to rebound from its Century of

Competitor Approaches to Irregular Warfare   47

Humiliation.[20] One thing China and the US have in common is that both exacerbate their competition by asserting—with certainty bordering on hubris—that their system of government is the best.

Russian Strategic Culture
Russian strategic culture has made the Kremlin more aggressive than China, less risk-averse, and more willing to cross the threshold of overt military conflict in its attempts to counter US and Western influence. Indeed, "Russia is a rogue, not a peer: China is a peer, not a rogue."[21] Thus, although Russia's strategic culture inclines toward IW, it presents different challenges.

Like the US and China, Russia's history and geography also profoundly influence its strategic culture. Due to its geographical vastness, maintaining territorial control, preventing the rise of separatist movements, and securing its borders represent enduring challenges in Russian history. In particular, Russia's vulnerability to brutal invasions[22] like those by Napoleon and Hitler has produced a near obsession with defense and security, and successive regimes in Moscow have since aimed to enhance the strategic depth on its frontier.[23]

Russia's history also includes several examples of both real and perceived foreign-sponsored subversion. The most prominent and proximate was the disintegration of the Warsaw Pact in 1989, followed by the collapse of the Soviet Union in 1991. Much like the CCP, Putin and his inner circle hold a deep-seated belief that the IW efforts of the US and its North Atlantic Treaty Organization (NATO) allies directly caused the collapse of the Soviet Union and subsequent democratization in Eastern Europe.[24] This has informed Putin's adoption of IW against the West to proactively counter what he perceives to be Western IW campaigns.[25]

A second pillar is Russia's strong central authority, from the tsars of the past to communist leaders to the present. However, there is no longer a strong party apparatus. Putin has built an authoritarian system that leverages personal relationships, transactional benefits, and a clan-like structure drawn largely from the second and third echelons of Soviet elites and oligarchs. This is tied together by self-interest, conservative nationalism, and a paradoxical ability to garner support from ethnic minorities.[26] Russia's authoritarian compartmentalization is effective in preventing coups and supporting covert IW efforts to provide plausible deniability. However, it also causes an overreliance on state security services that hinders the overall coordination of IW efforts.

The final critical pillar of Russia's strategic culture is nostalgia for, and administrative ties to, the former Soviet Union. Though Russia's leaders may reject Soviet communist orthodoxy, they are well-versed in Cold War IW methods. This includes a tradition of employing "active measures" or covert political operations ranging from disinformation campaigns to staging rebellions.[27] Such activities date to tsarist efforts to build false narratives to gain political support, which the Soviet *Komitet Gosudarstvennoy Bezopasnosti* (KGB), its primary intelligence agency, perfected and institutionalized.[28] But this challenge is even more dangerous today; modern tools like social media have increased the potency of such active measures beyond the wildest dreams of both the tsars and the KGB.

Russian and American strategic cultures differ in three major ways. First, Russian leaders reject the distinction between war and peace and are more risk-acceptant of undertaking military operations to defend territory perceived as Russian and to garner regional influence. Second, Russia's motivations are tied to nationalism and Russian (or greater Russian) identity. Third, Russia invests its limited resources primarily in neighboring countries and wherever it believes it needs to build the power, influence, and legitimacy to maintain its status as a world power. But it is also ready to seize key opportunities far from home where it sees a gap in US or Western influences.[29] The following chart captures each competitor's strategic culture and their respective use of IW tools.

Competitor Approaches to Irregular Warfare 49

Figure 4. Competitor Strategic Culture and Use of IW Tools.

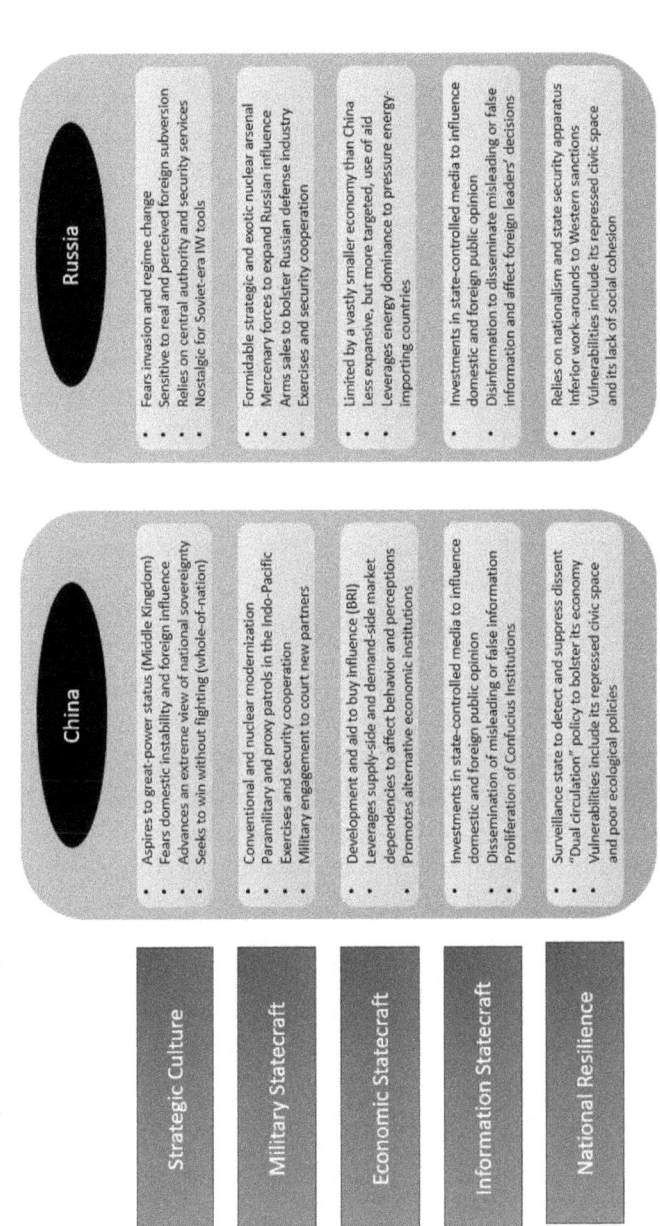

Strategic Culture

China
- Aspires to great-power status (Middle Kingdom)
- Fears domestic instability and foreign influence
- Advances an extreme view of national sovereignty
- Seeks to win without fighting (whole-of-nation)

Russia
- Fears invasion and regime change
- Sensitive to real and perceived foreign subversion
- Relies on central authority and security services
- Nostalgic for Soviet-era IW tools

Military Statecraft

China
- Conventional and nuclear modernization
- Paramilitary and proxy patrols in the Indo-Pacific
- Exercises and security cooperation
- Military engagement to court new partners

Russia
- Formidable strategic and exotic nuclear arsenal
- Mercenary forces to expand Russian influence
- Arms sales to bolster Russian defense industry
- Exercises and security cooperation

Economic Statecraft

China
- Development and aid to buy influence (BRI)
- Leverages supply-side and demand-side market dependencies to affect behavior and perceptions
- Promotes alternative economic institutions

Russia
- Limited by a vastly smaller economy than China
- Less expansive, but more targeted, use of aid
- Leverages energy dominance to pressure energy-importing countries

Information Statecraft

China
- Investments in state-controlled media to influence domestic and foreign public opinion
- Dissemination of misleading or false information
- Proliferation of Confucius Institutions

Russia
- Investments in state-controlled media to influence domestic and foreign public opinion
- Disinformation to disseminate misleading or false information and affect foreign leaders' decisions

National Resilience

China
- Surveillance state to detect and suppress dissent
- "Dual circulation" policy to bolster its economy
- Vulnerabilities include its repressed civic space and poor ecological policies

Russia
- Relies on nationalism and state security apparatus
- Inferior work-arounds to Western sanctions
- Vulnerabilities include its repressed civic space and its lack of social cohesion

Competitor IW Approaches

Drawing upon this book' division of IW tools into four categories—military statecraft, economic statecraft, information statecraft, and resilience—this chapter next applies this framework to competitor approaches.

Chinese Military Statecraft

As China continues to build its military power and improve its joint integration for warfighting, it also wields the tools of military statecraft to build influence and legitimacy worldwide. The CCP's holistic view of competition envisions military force as more than a coercive instrument. If war is politics by other means, then political-military action can be considered war by other means.[30] Therefore, military statecraft is part of a broader foreign policy effort to create a favorable international image, increase influence, collect intelligence, learn from advanced militaries, and shape the security environment in China's favor.[31]

Over the past two decades, China has increased its use of military statecraft, with multiple factors driving this expansion. First, the People's Liberation Army (PLA) has improved its capabilities, generating more options and growing confidence in their use. In addition to its impressive conventional and nuclear modernization, China has also leveraged paramilitary, proxy, and covert efforts to expand its influence while maintaining a veneer of legitimacy. For example, to advance revisionist territorial claims in the South and East China Seas,[32] Beijing has the People's Armed Forces Maritime Militia (PAFMM), "a so-called maritime militia of well-equipped vessels numbering in the hundreds—disguised as fishing vessels that patrol, surveil, resupply, and sometimes provoke."[33] This militia has expanded Chinese force posture in the region, with Beijing threatening and detaining fishermen of other nationalities under the guise of "policing."[34] China has combined these covert and proxy activities with conventional military means to signal resolve and project power. This includes deploying its aircraft carriers, conducting joint, cross-domain military exercises, landing nuclear-capable bombers on

Competitor Approaches to Irregular Warfare 51

contested islands, and constructing and militarizing artificial islands.[35] The integration of Chinese conventional and paramilitary forces demonstrates their increasing sophistication.[36]

Second, China's rising international engagement, particularly through its Belt and Road Initiative (BRI), has provided greater global access and opportunities for security cooperation. Beijing has increasingly relied on private military and security companies (PMSCs) to work alongside BRI projects. Although estimates vary, China may have as many as 40 different PMSCs operating abroad in 40 countries and more than 7,000 such companies domestically.[37] The PLA has also expanded its military exercises beyond the Indo-Pacific, most notably into Central Asia, the Middle East, and Africa, to conduct anti-piracy and related operations in support of BRI countries.[38] These engagements allow China to expand its influence, secure access to resources, and facilitate greater familiarization with new military capabilities.

Finally, under Xi, the CCP has encouraged the use of military capacity to reinforce diplomacy. In particular, China has employed military engagements to court new partners in Asia, the Middle East, and Africa.[39] Phillip C. Saunders and Jiunwei Shyy document a 27% growth in PLA interactions with international partners from 2002 to 2018.[40] The real change lies in the composition of these engagements. The PLA has shifted its emphasis from senior-level visits to more robust military exercises and port calls, emphasizing combat and combat support operations with Shanghai Cooperation Organization members[41] and new BRI partners.

Like the US, China also orchestrates international professional military education programs to train officers of other countries[42] and uses arms sales and military aid to bolster its partners' capabilities and interoperability to expand the prestige of its own growing defense industry.[43] The establishment of common training and supply chains imbues China with coercive power and influence over the increasingly dependent political and military leaders of these partner countries.[44] Lastly, the PLA has participated in humanitarian, disaster relief, and UN peace operations, lending further legitimacy to Chinese military statecraft.

Russian Military Statecraft

The Kremlin maintains the Soviet tradition of employing military statecraft extensively to increase power, gain influence in client states, and, ultimately, counter Western influence. Though Russia lacks the relative military capacity of the Soviet Union, it has engaged in targeted use of military statecraft. Russia exploits a variety of conventional, nuclear, and paramilitary tools to bolster its military power vis-à-vis the US and NATO.

Despite the military modernization in the early years of Putin's presidency, the Russo-Ukrainian War demonstrated that these reforms did not go far enough to overcome joint planning and logistical weaknesses. The war will likely render the Russian military a shadow of its former self, and another conventional invasion in Eastern Europe will be less likely for the foreseeable future. Nonetheless, Moscow has a formidable arsenal of strategic, nonstrategic, and exotic nuclear weapons, as well as other technologically advanced capabilities like air and missile defenses and hypersonic weapons.[45]

Moreover, paramilitary and proxy forces serve as force multipliers to expand Russian influence while maintaining plausible deniability. For example, the Kremlin has supported ethnic Russian separatist movements in Moldova, Ukraine, and Georgia, while collaborating with a variety of terrorist groups and nonstate actors in the Middle East.[46] The most notorious paramilitary force has been the Wagner Group—a PMSC that has conducted combat operations and military training to support friendly regimes, secure access to resources, build influence and legitimacy, and undermine Russia's enemies.[47] But the weakness of this approach became evident in July 2023, when Wagner leader Yevgeny Prigozhin—believing that the Kremlin did not properly equip or respect his mercenaries—led a brief rebellion, which likely resulted in his subsequent death in a plane crash a short time later.[48] The episode serves as a reminder that IW tools, though often helpful, can backfire.[49]

Like the Soviet Union, the current Russian regime uses military tools to support friends and undermine Western influence, primarily through arms sales.[50] Due to both legacy sales and Putin's modernization program, the Russian defense industry maintains some of its former prestige. Despite its failures in Ukraine, Russia remains the second largest arms exporter behind the US—accounting for just under twenty percent of global sales and generating $15 billion in revenue.[51] These exports provide advanced weaponry to various countries, often at a lower cost than US systems, courting new allies and partners in the Middle East, Africa, and Asia.[52] Russian military statecraft even threatened to cleave Turkey from the rest of NATO in 2017 due to concerns about Turkey's purchase of S-400 systems.[53] But there is growing evidence that Russian arms sales will continue to decrease because of the war it waged on Ukraine,[54] thereby presenting opportunities for the US.

Russia has used multiple means of security cooperation to expand its influence. Russia conducts joint military exercises with various allies and partners as a means of garnering geopolitical influence, demonstrating military capabilities, and advancing an alternative military bloc.[55] Over the last decade, Russia has aggressively sought security cooperation agreements around the world particularly in Africa and the Middle East, signing over two dozen such agreements to allow varying levels of Russian access and cooperation.[56] But since its 2022 invasion of Ukraine, Russia has focused on greater military cooperation with China, Iran, and even North Korea to supply critical weapons and materiel for its war effort.[57]

Chinese Economic Statecraft
China employs economic statecraft to wield power, build influence, and increase its legitimacy as a global leader. After joining the World Trade Organization (WTO) in 2001, China's Gross Domestic Product (GDP) rose

from 10% to 78% of US GDP in 2021.[58] Over this period, China developed a sophisticated economic toolset rivaling that of the US.[59] Rather than making China more liberal or cooperative as Western powers had hoped, this has provided Beijing with the instruments, influence, and credibility to challenge the US-led order.

Beijing employs development support and financial assistance to enhance its influence and legitimacy. Through BRI and its leadership of the Asian Infrastructure Investment Bank, China has facilitated over a trillion dollars in investment in infrastructure worldwide, with a long list of countries in its networks.[60] Though Western-led development efforts and institutions are still larger in membership and loans over time, Chinese alternatives are often attractive due to their less stringent requirements and conditions.[61] However, global opinion has begun to sour on these institutions due to their lack of transparency, accountability, and proclivity for locking partner states into overextension.[62]

Beijing also facilitates such dependencies by leveraging its dominant position in international trade and finance. By controlling crucial production sectors and wielding access to its huge consumer markets, China exerts pressure on governments to comply with its geopolitical preferences. Targeted sanctions and restrictions are particularly potent against countries that are critically dependent on Chinese supply chains and markets, thereby allowing China to accrue greater power and influence to affect behavior and perceptions. However, more robust economies can offset such dependencies by relying upon alternatives.[63]

Recognizing the power, influence, and legitimacy of US-led Bretton Woods economic institutions, China works with other states—including Russia—to create and expand competing regional institutions. For example, they have tried to strengthen relations with fellow BRICS countries (Brazil, India, and South Africa) and have developed rival trade and financial networks.[64] Moreover, China—with Russia and other competitors' support—attempts to reduce reliance on the US dollar and challenge its status as the dominant global currency. Accordingly, Beijing seeks to promote its own national currency, the Renminbi (RMB), along with other currencies, by encouraging bilateral trade agreements that bypass the US dollar and fostering currency swap arrangements that make it easier for countries to conduct business in their own currencies.

China also works with partners to establish alternative international payment systems to the US-dominated Society for Worldwide Interbank Financial Telecommunication network (SWIFT). Beijing has already developed the Cross-Border Interbank Payment System, allowing direct RMB transactions between participating countries. These competing institutions threaten to alter the flow of global financial transactions and bolster Chinese legitimacy among various actors who chafe under US economic leadership. Yet, whether they can overpower the inertia and longstanding geopolitical influence of the US dollar and the entrenched institutions that depend on it remains uncertain.[65]

Russian Economic Statecraft
Although Moscow's relatively smaller economy (11th in the world nominally as of 2023) prohibits it from wielding Beijing's same economic toolset,[66] it has capably employed key economic dependencies as part of broader IW campaigns. Like China and the US, Russia uses development assistance and foreign aid to restore relationships with former Soviet republics, support friendly regimes worldwide, and invest in strategic projects to build a loose coalition of relatively dependent states willing to oppose the US-led order. Specific initiatives include economic aid, technical assistance, and infrastructure projects that rely on longstanding Russian expertise in areas like nuclear power generation.

The scale of such Russian assistance is small—their own reporting listed it as $902 million in 2015—compared to the $39.2 billion the US provides in annual aid. The Kremlin maximizes its effectiveness by targeting countries that provide valuable resources or share strong anti-US sentiment.[67] For example, in July 2023, at the Africa Summit in St.

Petersburg, Putin promised grain shipments to Mali, Burkina Faso, and Central African Republic, among others, to curry influence.[68]

Debt forgiveness is another tool Russia uses to gain influence and legitimacy in Africa; Russia recently signed an agreement to forgive Somalia's $684 million in outstanding debt.[69] Russia also supports the efforts of China and other BRICS to reduce dependence on the US dollar and establish competing institutions. Russia's alternative to the SWIFT system, its System for Transfer of Financial Messages, has grown in the wake of the Russo-Ukrainian War and the unprecedented sanctions levied against it,[70] demonstrating the risks of US overreliance on economic sanctions as an IW tool (as discussed in chapter 6).

Russia's most effective economic tool has been its energy dominance. In particular, Moscow leverages dependence on Russian energy resources to "cement political alliances and drive wedges within counter-alliances."[71] Accordingly, Russia often employs price manipulation and selective supply cuts to pressure energy-importing European countries. Researchers estimate that direct or threatened use of such tools has occurred over 50 times.[72] However, as a reminder of the risks of overusing IW tools, the repeated use of this tool has encouraged some states, such as Lithuania and Poland, to develop energy alternatives to mitigate Russia's use of energy as a weapon of war.[73]

Chinese Information Statecraft

Autocratic regimes like China enjoy certain advantages in the control of information for strategic purposes (see chapter 7). The CCP conceives of information statecraft through the framework of its Three Warfares strategy, which comprises:

- "public opinion warfare" to influence the perceptions of domestic and foreign publics;
- "psychological warfare" to build or exert influence over foreign elected leaders, opposition groups, or military leaders, affecting the execution of their policies vis-à-vis China;

- "legal warfare" to bolster legitimacy by constructing legal justifications for Chinese actions and promoting alternative interpretations of international law.[74]

Though often portrayed as PLA military doctrine, the Three Warfares are fundamental responsibilities of the CCP, which coordinates their use domestically and globally.[75]

In conducting public opinion warfare, China has made unprecedented investments to control domestic public opinion and shape perceptions abroad. Internationally, Beijing has funded its own media enterprises to reduce Western influence. For example, China Global Television Network has a $10 billion annual budget to spread the CCP's narratives on current events, compared to the US Agency for Global Media's (the US government's comparator) paltry $637 million budget that funds objective media outlets.[76] Additionally, the CCP supports the establishment of Confucius Institutes, which have enrolled almost 9 million students in over 160 countries.[77] Although ostensibly created to provide Chinese language instruction, the centers have become notorious for pushing an ideological agenda, limiting free speech, and attempting to silence anti-PRC debates.[78] Although these institutes have declined in popularity in many Western countries, they remain popular in Africa, Latin America, and the Middle East.[79]

Disinformation is a critical pillar of the Three Warfares. The CCP and its sponsored organizations, including private companies to which campaigns are outsourced,[80] manipulate public opinion in target countries by disseminating misleading or false information (public opinion warfare).[81] This tactic complicates the decision-making environment of foreign leaders (psychological warfare) and counters criticism on issues like human rights and territorial disputes (legal warfare). The means for co-opting information are often both overt (through investments in media outlets) and covert (through financial and other support for high-profile political and media leaders).[82] Thus, information statecraft has enabled China to shape a global narrative that aligns with its interests while undermining those of Western democracies.

Russian Information Statecraft

Like China, Russia has attempted to spread favorable narratives through state-controlled media. For example, RT and Sputnik (Russian state-directed and funded media outlets) influence foreign audiences by disseminating biased news, promoting divisive content, supporting pro-Russian political figures, undermining trust in Western institutions, and targeting Russian-speaking communities.[83] The Kremlin also promotes Russian language, history, and values through "Russian Cultural Centers" and the Russkiy Mir (Russian World) foundation, especially targeting neighboring states with existing cultural affinities or large ethnic Russian populations.[84]

Russia is notorious for its extensive information operations to spread disinformation. Covertly, the Kremlin leverages fake social media accounts and bots—while co-opting political and media leaders—to promote the narratives that state media outlets overtly advance. The aim is to exacerbate political divisions in Western countries. A prominent example is Russian interference in the 2016 US presidential election, which polarized voters and fostered mistrust in US political processes. Russia also uses similar tactics to bolster far-right political parties in Europe, such as Italy's Lega Party and Austria's Freedom Party, aggravating divisions between the US and its allies.[85] Russian information operations are supported by a vast network of both state actors and non-state proxies, including military intelligence organizations like the Glavnoye Razvedyvatel'noye Upravleniye (GRU) and Sluzhba Vneshney Razvedki (SVR), as well as hackers, business and economic interests, and even the Russian Orthodox Church.[86]

Resilience

In an era of crises, resilience has become a source of power. The Fund for Peace developed its State Resilience Index (SRI) to measure national resilience, defining it as "the extent to which a country can prepare, manage, and recover from a crisis."[87] The SRI classifies Russia and China as moderately resilient. China's civic space, environment, and ecology are its biggest weaknesses, whereas individual capabilities and social cohesion are strengths. Russia's civic space and social cohesion are its biggest weaknesses, whereas individual capabilities and state capacity are relative strengths.

National Resilience—China

Xi Jinping has recently consolidated China's national resilience efforts under his Central National Security Committee.[88] Speaking about resilience at the 20th CCP Congress address in October 2022, Xi emphasized China's efforts at "holistic national security." This concept defines national resilience in terms of a proactive party and government that ensures social stability, disaster prevention, emergency management, and economic stability.[89] The Chinese government also mentioned the importance of enhancing economic resilience in the 2022 Central Economic Work Force Conference.[90] To deepen its resilience, the CCP relies on surveillance and economic policies.

The CCP has long focused its national resilience efforts on building a formidable surveillance state. In the years since Tiananmen, the CCP spent billions on domestic security forces to detect and suppress even the smallest signals of dissent.[91] Though originally reliant on human informants, today China largely relies on AI-enabled surveillance to monitor and control information flows into and throughout China.[92] This surveillance, China's national internet firewall, and domestic information domination negatively impact civic space, weakening aspects of national resilience in the long run.[93]

China also uses economic planning, such as investments in critical technology,[94] to build its own resilience while generating coercive power. Some have referred to these policies as "dual circulation," through which China builds economic resilience by reducing dependence on foreign supply chains and increasing domestic production and consumption. This two-sided industrial policy ensures domestic resilience and other countries' dependence, weakening supplier or consumer resilience.[95]

National Resilience—Russia
Russia's national resilience relies on nationalism and a powerful Federalnaya Sluzhba Bezopasnosti (FSB)-led state apparatus. In the lead-up to its 2022 invasion of Ukraine, the Russian government took specific measures to improve national resilience in the political, information, and economic spheres. This included legal efforts to cement the regime's authority and leveraging Soviet nostalgia and narratives of the restoration of greater Russia. In the first weeks of the war, the Duma (Russia's legislative branch) passed a law threatening journalists who report "false information" about the Russian military with up to 15 years in prison. That forced two of the largest remaining independent Russian news sources to shut down (Dozhd TV) or stop all coverage of the war in Ukraine (Novaya Gazeta).[96]

The regime also reenergized efforts to limit the impact of anticipated Western economic sanctions.[97] For example, Russia increased its foreign exchange and gold reserves from $363 billion in 2015 to $643 billion[98] on the eve of the invasion, though half of these are now under sanctions and inaccessible.[99] The head of the Russian Orthodox Church, Patriarch Kirill, has declared Russia's invasion of Ukraine a holy war, bolstering domestic support for the war effort.[100]

CONCLUSION

The strategic cultures of China and Russia facilitate the ease of their use of IW approaches. China is particularly well-organized to employ IW tools, thereby maintaining some advantage over the US. Russia has had more limited successes, and the conflict in Ukraine is reducing Russia's relative power, influence, and legitimacy. The openness of Western societies

Competitor Approaches to Irregular Warfare   61

presents Russia and China with some advantages, such as susceptibility to disinformation campaigns. Conversely, other elements of democratic societies, particularly economic institutions and the vast ecosystem of journalism and media outlets, put Russia and China at a disadvantage. Regardless, without concerted reforms and efforts to bolster resilience, the US and Western democracies will remain vulnerable to competitors' IW strategies.

However, China's and Russia's overreliance on IW has resulted in some diminishing returns for their power, influence, and legitimacy and has even backfired occasionally.[101] Russian IW approaches, especially those used in conjunction with the conventionally executed Russo-Ukrainian War, have mobilized robust international responses: unprecedented economic sanctions, a renewed sensitivity to disinformation, economic coercion, and election meddling. Ironically, their actions have resulted in a renewed commitment to NATO.[102] China has received similar backlash, as evidenced by falling influence and favorability indicators.[103]

A second weakness is the inherent fragility of authoritarian political systems. Although China and Russia have undertaken concerted efforts to protect their populations from foreign influence, their approaches create vulnerabilities whenever narratives fail to match reality. The Wagner rebellion demonstrates the danger in creating parallel military structures that rival state institutions, which may ultimately generate regime vulnerability. The Kremlin granted the WagneGroup significant power and autonomy to preserve its plausible deniability—leeway that eventually led to a failed insurrection. China's weaknesses—the inflexibility of the zero-COVID policy and its subsequent economic and social crises—were exposed to the world in late 2022, prompting some demonstrators to display blank sheets of paper to represent all they could not say.[104] Indeed, "China's political economy system is robust, but it is not resilient."[105]

Ultimately, this chapter exposes a great irony. While the US struggles to define and implement IW due to a strategic culture that is diametrically opposed to it, Chinese and Russian strategic cultures integrate IW into their narratives, bureaucracies, and approaches. Yet, it is China and Russia's leaders who most fear American IW. They unequivocally believe in its efficacy, recognize that they will always be vulnerable to it, blame it for their greatest strategic defeats, and seek to preempt it. Therefore, rather than continuing to marginalize such approaches, the US and its allies should embrace IW as a means of increasing power, influence, and legitimacy. As chapter 4 illustrates, the US successful use of IW—specifically political warfare—against historical adversaries demonstrates that Washington can and should reinvest in these capabilities.

Notes

1. McFate, "Irregular warfare with China, Russia."
2. Jones, *Three Dangerous Men*.
3. Thomas, *The Chinese Way of War*.
4. Fearon, "Signaling Foreign Policy Interests"; and Gleiman, *The Unconventional Strategic Option*.
5. Reiter and Stam, *Democracies at War*.
6. Westad, "Legacies of the Past."
7. Wang, *Never Forget National Humiliation*.
8. Nathan and Scobell, "How China Sees America."
9. Liu, *The China Dream*.
10. Huang, *The Rise and Fall of the East*.
11. Ota, "Sun Tzu in Contemporary Chinese Strategy."
12. Jones, *Three Dangerous Men*, 133.
13. Nathan, "The Tiananmen Papers."
14. Zhang et al., *The Tiananmen Papers*.
15. US Department of Defense, *Military and Security Developments*.
16. Fravel, "China's Sovereignty Obsession."
17. R2P is a nonbinding norm holding that outside powers have a responsibility to intervene in response to a government's failure to protect its population from mass atrocities such as genocide, even though this violates the government's sovereignty. The world's governments adopted the R2P doctrine in 2005, but it remains controversial and unevenly applied. Simonovic, "The Responsibility to Protect."
18. This was the justification for intervention in Korea in 1951 and in Vietnam in 1979.
19. Kim and Prytherch, "Douzheng."
20. Johnson, "The 'Know Thyself' Conundrum," 122.
21. Dobbins, Shatz, and Wyne, "Russia Is a Rogue."
22. Kerrane, "Moscow's Strategic Culture."
23. Rumer and Sokolsky, "Etched in Stone."
24. Galeotti, "Active Measures"; Rumer and Sokolsky, "Etched in Stone"; Way, "The Real Causes"; and Bunce and Wolchik, *Defeating Authoritarian Leaders*.
25. Bartles, "Getting Gerasimov Right."

26. Graham, "The Precarious Future"; Blitt, "Russia's "Orthodox" Foreign Policy"; and Alexseev, "Russia's Ethnic Minorities."
27. Galeotti, "Active Measures."
28. Rid, *Active Measures*; the English translation for KGB (Komitet Gosudarstvennoy Bezopasnosti) is Committee for State Security.
29. German and Kuhrt, "Will Russia's Global Influence."
30. Babbage, *Winning Without Fighting*, 8; Clausewitz, *On War*, 87; and Singh, "Rereading Mao's Military Thinking."
31. Allen, Chen, and Saunders, *Chinese Military Diplomacy, 2003–2016*.
32. Council on Foreign Relations, "China's Maritime Disputes"; and Ngo and Koh, "Lessons from the Battle."
33. Nguyen et al., "China Is Winning."
34. US Department of Defense, *Military and Security Developments*.
35. US Department of Defense; and CSIS, Asia Maritime Transparency Initiative, "China Island Tracker."
36. Crouch, Pavel, Starling, and Trotti, "A New Strategy."
37. Markusen, "A Stealth Industry."
38. Standish, "Central Asia Caught."
39. Saunders and Shyy, "China's Military Diplomacy."
40. Saunders and Shyy.
41. Wolfley, *Military Statecraft and the Rise*, 30.
42. Marat, "China's Expanding Military Education"; and Nantulya, "Chinese Professional Military Education."
43. CSIS, ChinaPower, "How Dominant Is China?"
44. Seldin, "China's Speed in Selling Arms."
45. Kroenig, Massa, and Trotti, "Russia's Exotic Nuclear Weapons."
46. Jones, "The Future of Competition."
47. Rampe, "What Is Russia's Wagner Group Doing in Africa?"
48. Stewart and Ali, "US Asks What's Next for Wagner Group."
49. Aydıntaşbaş et al., "What is the fallout of Russia's Wagner rebellion?"
50. Krause, "Military Statecraft," 315.
51. Green, "Russian Arms Exports."
52. Bergmann, Snegovaya, Dolbaia, and Fenton, "Seller's Remorse."
53. Sanders and Akca, "The Great Unwinding."
54. RFE/RFL Staff, "Russia's Global Arms Exports."
55. Wolfley, "Military Statecraft and the Use."
56. Siegle, "Russia and Africa."
57. Britzky, "Top US General Says.'"
58. Allison, Kiersznowski, and Fitzek, "The Great Economic Rivalry."

59. Weiss, "Re-Emergence of Great Power Conflict."
60. Wang, "China Belt and Road Initiative (BRI) Investment Report 2022"; and Asian Infrastructure Investment Bank, *AIIB Report 2021*.
61. Asian Infrastructure Investment Bank, *Moonshots for the Emerging World*, "Introduction."
62. Himmer and Rod, "Chinese Debt Trap Diplomacy."
63. Leonard, "China Is Ready"; Economy, *The Third Revolution*; and Friedberg, "Globalisation and Chinese Grand Strategy."
64. Sarkar, "Trends in Global Finance."
65. Jin, "Why China's CIPS Matters (and Not for the Reasons You Think)."
66. Scholivin and Wigell, "Geoeconomic Power Politics"; and Zarate, *Treasury's War*.
67. Asmus, Fuchs, and Müller, "Russia's Foreign Aid Re-Emerges."
68. Behnke, "Putin searches for more friends."
69. "Somalia says Russia grants relief on debt worth $684 million."
70. Marrow, "Russia's SWIFT Alternative Expanding."
71. Wigell and Antto, "Geopolitics versus Geoeconomics"; and Wigell, Scholvin, and Aaltola, *Geo-Economics and Power Politics*.
72. Larsson, *Russia's Energy Policy*.
73. LaBelle, "Energy as a Weapon of War."
74. Mattis, "China's 'Three Warfares' in Perspective."
75. Lee, *Chinese Political Warfare*; and Mattis, "Form and Function."
76. Gupta, "Panda Power?" These outlets include Voice of America, Radio Liberty, and Radio Free Asia.
77. Chai, "China's Nascent Soft Power Projection."
78. Chai.
79. Peterson, "China's Confucius Institutes."
80. Jacobs and Carley, "#WhatIsDemocracy."
81. Harold, Beauchamp-Mustafaga, and Hornung, *Chinese Disinformation*.
82. Ni, "China Funnels Its Overseas Aid."
83. Paul and Matthews, "The Russian 'Firehose of Falsehood.'"
84. Efremenko, Ponamareva, and Nikulichev, "Russia's Semi-Soft Power."
85. Jones, "The Future of Competition."
86. Zakem, Saunders, and Antoun, "Mobilizing Compatriots."
87. Fund for Peace, "State Resilience Index."
88. Drinhausen and Legarda, "'Comprehensive National Security.'"
89. Wishnick, "Xi Jinping's Vision."
90. Li and Huang, "The Latest Research Progress."
91. "China Spends More on Controlling Its 1.4bn People than on Defense."

92. Peterson, "Designing Alternatives."
93. Friedberg, "Globalisation and Chinese Grand Strategy."
94. US Department of Defense, *Military and Security Developments.*
95. Leonard, "China Is Ready"; Economy, *The Third Revolution*; and Friedberg, "Globalisation and Chinese Grand Strategy."
96. "Russia fights back"; and Kuznetsova, "Putin Shuts Down Russia's Free Press."
97. Stoicescu, Nazarov, Giles, and Johnson , "How Russia Went to War."
98. "International Reserves of the Russian Federation (End of period)."
99. Nelson et al., *Russia's War on Ukraine.*
100. "Russian Patriarch Kirill Says."
101. Kim, "Why China's Propaganda Efforts"; and Ma, "Pew Survey: Global Opinion of China."
102. US Department of Treasury, "Treasury Sanctions Russian Intelligence Officers"; and Taylor, "Combating Disinformation."
103. Linetsky, "China Can't Catch a Break."
104. Perrigo, "Why a Blank Sheet."
105. Jin, *The New China Playbook.*

Chapter 4

The Past as Potential Prologue: Political Warfare during the Cold War

> History never repeats itself, but it does often rhyme.[1]
> Attributed to Mark Twain

Introduction

As the United States considers a grand strategy for the "decisive decade,"[2] it is helpful to examine its Cold War experience with political warfare. Such an evaluation provides insights as to its utility for the current era of competition and crises, as well as its legitimacy in the eyes of the American public and its partners and allies. Although contemporary circumstances differ significantly from those of the Cold War, there are enough similarities to warrant a comparative analysis without assuming the same approaches that worked over half a century ago will work today.

The parallels between the two eras are substantial. They include long-term competition for global primacy with a near-peer competitor, an underpinning of ideological differences, multimodal forms of competition on a global scale that employ all instruments of power, and a desire by all major powers to avoid direct conventional military conflict, coupled with a demonstrated willingness to engage in multiple forms of asymmetric conflict.

To use the phrase attributed to Mark Twain, although the parallels between today and the Cold War may "rhyme," they are far from an exact repetition. First, although the current era is marked by ideological differences, these differences are not as profound as they were during the Cold War. Although nominally professing communism, China has instead pursued aggressive economic development through capitalism tightly controlled by the CCP. Second, the current international system is evolving toward multipolarity.[3] Russia continues to play a significant role in the international system, as do other middle powers. Third, economic interdependence has substantially changed the playing field. Although globalization has not led to widespread democratization as some hoped it would in the 1990s,[4] it has changed the geopolitical environment. American trade with the Soviet Union was minuscule compared to that of the US and China today.[5]

Furthermore, there are dependencies in both directions. The US is currently dependent on China for rare earth elements necessary for critical electronics, although the Biden administration is working to develop alternate sources of supply.[6] China is currently at least 70% dependent on the US and its allies for more than 400 products running the gamut from luxury items to raw materials.[7] Thus, economic interdependence complicates geopolitics in ways that cannot be ignored. Finally, the world is entering an era of crises characterized by climate effects, disease outbreaks, and cyberattacks[8] in tandem with strategic competition.[9] Despite these differences, the parallels between the two eras remain

The Past as Potential Prologue

significant enough to merit an investigation of political warfare techniques used during the Cold War.

This chapter first examines Kennan's definition of political warfare and the major factors that influenced his approach. It also considers the term's placement within the taxonomy of IW and explores political warfare's relationship to America's Cold War containment strategy, with an eye toward the concept's contemporary applicability. It then analyzes the recent resurgence of interest in the concept, including arguments promoting political warfare as a basis for contemporary American grand strategy.

POLITICAL WARFARE WITHIN THE BROADER TAXONOMY OF IRREGULAR WARFARE

This section first defines political warfare in relation to IW. It then explains how domestic politics impacted the development and implementation of political warfare ideas in the US. The multiple faces of political warfare are described to include its many forms through historical examples, chronicling its use from the outset of the Cold War through the Reagan Years, noting the pivotal role of US domestic public opinion. The section ends with an overview of the enduring lessons.

Political Warfare Defined

In a 1948 DOS Policy Planning Staff Memorandum, Kennan defined political warfare as:

> the logical application of Clausewitz's doctrine in time of peace. In broadest definition, political warfare is the employment of all the means at a nation's command, short of war, to achieve its national objectives. Such operations are both overt and covert. They range from such overt actions as political alliances, economic measures (as ERP[10]), and "white" propaganda to such covert operations as clandestine support of "friendly" foreign elements, "black"

psychological warfare and even encouragement of underground resistance in hostile states.[11]

This differs from Jones' IW definition in only its wording. Although Kennan's description is more detailed, their meanings are identical. Jones defines IW as "activities short of conventional and nuclear warfare that are designed to expand a country's influence and legitimacy, as well as to weaken its adversaries."[12] Thus, there are few activities these expansive definitions do not encompass short of direct military conflict.

Capitalizing on its breadth, the American government, then new to its role as a global leader, used the concept of political warfare to conceptualize security broadly. It widened its definition of conflict beyond strictly military confines to comport with the intent of the National Security Act of 1947.[13] Political warfare also provided an intellectual foundation for Kennan's Cold War containment strategy, although the definition of containment underwent multiple incarnations over its four-decade lifespan.[14]

Taking Domestic Politics into Account

There is a tendency among historians and strategists to regard the Cold War as the "golden age" of American strategic thinking. However, the truth is far messier and more complex, so much so that Mahnken refers to the "myth and metaphor of containment"—signifying the "simplified and idealized representation" of the strategy as that which is fondly remembered rather than the strategy itself.[15] Although there was an indisputable logic to containment, the concept evolved to encompass multiple ideas and approaches during the Cold War. What they all shared was a common enemy—the Soviet Union—and an ability to fit within Kennan's broad definition of political warfare.

However, the broadness of political warfare as a concept did not guarantee its coherence. Hal Brands opines that the term is "vague and unspecified."[16] Frank Hoffman goes so far as to call it "oxymoronic" because there is no actual warfare in political warfare, at least not in the

The Past as Potential Prologue 71

Clausewitzian sense.[17] Scott Lucas and Kaeten Mistry argue that "in and of itself, political warfare did not define US strategy during the Cold War, nor did Kennan's musings on the approach prevent uncoordinated planning and implementation—both overt and covert."[18] Thus, the integration of the instruments of powers into a coherent strategy using political warfare as its basis was never fully actualized, which does not negate the fact that it was a core facet of America's Cold War strategy.

This conceptual vagueness is not shocking. In 1945, the US stepped into the role of global leadership, bureaucratically and temperamentally unprepared to assume the mantle. Although the National Security Act of 1947 gave the US a structure around which to develop and implement a grand strategy, the passage of the legislation itself was an act of substantial political wrangling. Despite these significant obstacles, Kennan's concept of political warfare gave the Truman administration and the fledgling national security enterprise a flexible concept around which to operationalize American grand strategy.

At the end of World War II, the American public was anxious to return to its prewar way of life. Despite the plans by the US Navy and War Department for an orderly postwar demobilization of service members, Congress and the American public pressured the Truman administration for a demobilization that would be as rapid as possible. As a result, between September and December of 1945, the US Army demobilized more than 1.2 million soldiers per month.[19] Not to be outdone, the US Navy demobilized more than 3.1 million sailors, soldiers, and marines from the Pacific during the second half of 1945.[20] By the official demobilization completion on June 30, 1947, the US Army's end strength decreased from a wartime high of more than 8 million soldiers to merely 684,000.[21] The other services suffered similar reductions.

Decreases in military funding accompanied this rapid decline in available manpower. President Harry S. Truman, determined to balance the national budget and erase the accumulated wartime deficits, developed and employed the "remainder method" for determining military expenditures.[22] This method calculated the DOD budget from the leftover funds after meeting all other national requirements. The administration began using this calculus in 1946, abandoning it only after the 1950 outbreak of the Korean War.[23]

The post–World War II military drawdown provides perspective on the circumstances Kennan faced when he coined the term "political warfare." Not only were there fewer forces to project military power, hence the need for other instruments; the abysmal performance of Task Force Smith in 1950, during which the American reinforcements were nearly expelled from the continent by the invading North Korean Army, demonstrated the diminished readiness in those who remained on active duty.[24] Given this, the turn towards political warfare—an asymmetric approach integrating all instruments of national power—makes sense. Kennan was using the means available to strategic advantage.

The US currently faces similar circumstances: a multiplicity of required security investments it cannot afford. These range from hardening ports against rising seas to providing greater protection against growing cyber threats to repairing crumbling critical infrastructure while maintaining investment in cutting-edge military technology. Therefore, the Cold War containment strategy, underpinned by the significant use of political warfare, is a reasonable starting point for developing an American grand strategy for the twenty-first century.

The Multiple Faces of Political Warfare

During the Cold War, political warfare took many forms. Given the breadth of Kennan's definition, it is helpful to provide some foundational attributes of the term and examples of its use. Danny Pronk lists the following fundamental characteristics:

- Employs all the elements of national power
- Relies heavily on unattributed forces and means (e.g., covert operations)
- Stays below the legal threshold of open, armed conflict

The Past as Potential Prologue 73

- Can achieve effects at lower costs
- Exploits shared bonds (ethnic or religious) or seams
- Requires a heavy investment in intelligence resources to detect[25]

This list illustrates how a state seeking to husband its resources would be drawn towards political warfare. Throughout the Cold War, America waged political warfare against the Soviet Union in multiple ways. It backed non-communist political parties in Europe and Asia, funded dissident activities behind the Iron Curtain, supported anti-communist artists and writers, and formed alliances whose purpose was to contain the spread of communism.[26]

Max Boot and Michael Doran explain that the American use of political warfare necessitated the establishment of a bureaucracy to manage it.[27] In addition to the development of a Central Intelligence Agency directorate to oversee covert operations, new organizations included Radio Free Europe, Radio Liberty, the US Information Agency, as well as the National Endowment for Democracy.[28] All were critical in supporting the Solidarity movement in Poland in the 1980s.[29] Even the Marshall Plan can be defined as an act of political warfare. Its purpose—beyond providing economic and humanitarian assistance—was to shore up Western Europe, making it more resilient and less vulnerable to acts of political warfare by the Soviets.[30]

Political Warfare at the Outset of the Cold War
John Lewis Gaddis argues the implementation of containment shifted between two poles during the Cold War: symmetrical and asymmetrical.[31] D. Robert Worley notes:

> Those presidents who believed their means were limited tended towards *asymmetric* responses to Soviet encroachments, that is, to select the place, time, magnitude, and methods of competition rather than responding tit-for-tat. Presidents who believed the American economy could produce the necessary means on

demand tended toward *symmetric* responses, countering Soviet adventurism wherever and whenever it occurred.[32]

President Truman began with asymmetric responses, only switching to a symmetric formulation when confronted with the Korean War.

In 1947, the Truman Doctrine declared that "It must be the policy of the US to support free peoples who are resisting attempted subjugation by armed minorities or by outside pressures."[33] This appears to have been a sweeping US commitment to defend democracy anywhere in the world and at any price. However, the reality underpinning it was far more circumscribed.[34] Gaddis argues that between 1947 and 1950, the Truman administration "...had neither the intention nor the capability of policing the rest of the world...."[35] Rather, the US quietly maintained a policy of bolstering Western Europe, containing the Soviet Union largely through economic measures.[36]

The Truman Doctrine also represented the first time Congress appropriated money to fight communism abroad.[37] Aid for Greece and Turkey grew into the far more ambitious European Recovery (Marshall) Plan. This was followed by the Berlin Airlift in 1948 and the establishment of NATO in 1949. Each was part of an integrated strategy of containment and constituted discrete acts in a political warfare campaign.

Those familiar with NATO will point to Article 5 of its Charter, which requires members to come to each other's aid in the event of an attack, as the alliance's most critical clause. However, the far less well-known Article 2 of the treaty is highly instructive when considering NATO's role in a political warfare campaign. It reads:

> The Parties will contribute toward the further development of peaceful and friendly international relations by strengthening their free institutions.... They will seek to eliminate conflict in their international economic policies and will encourage economic collaboration between any or all of them.[38]

Although the term "political warfare" naturally suggests offensive operations designed to weaken an enemy, it is critical to recall that Kennan's definition leaves room for defensive operations.[39] Thus, although offensive measures such as covert operations are certainly a tool of political warfare, so too are such measures as free trade, strengthening democratic institutions, creating political and military alliances, as well as strengthening domestic capacity to withstand shocks and crises. Therefore, Article 2 represents the further operationalization of an integrated political warfare campaign to contain the Soviet Union.

Courting Domestic Public Opinion
Against all odds, the administration sought and received the buy-in of the American people, at the time war-weary and wishing to disengage from international politics. Truman's strategy required convincing the public that the Soviet threat was so severe as to warrant America's continued defense of the "free world." This was accomplished through a deliberate information campaign.[40]

This was not simple. Bringing the American people on board required a concerted information campaign. Although the Marshall Plan eventually passed Congress, it required nearly a year of intense deliberation and lobbying and relied in large part on the apolitical reputation of Secretary George Marshall to forge a bipartisan coalition.[41] Congressional debate over the Marshall Plan, which ran from January to June 1948, saw arguments for and against on both sides of the aisle. These ranged from concerns over cost and feasibility to outright accusations of American imperialism.[42] At the outset, the American public was generally ignorant and apathetic regarding the plan.[43]

With the administration's blessing, the DOS began an extensive public relations campaign with incredible results. In 1947, the Committee for the Marshall Plan was formed to "stir up grassroots sentiment...in Congressional home districts."[44] They enlisted the assistance of local foreign affairs committees, chambers of commerce, and farming organizations.

Committee members wrote editorials, provided detailed maps of Europe's most affected areas (hoping to resonate with immigrants), and organized speakers with extensive radio coverage to tout the plan's virtues.[45] As a result, in merely six months, favorable opinion of the plan jumped from 56% to as high as 95%, with the number of voters familiar with the plan tripling during the same time frame.[46] Some analysts go so far as to credit Truman's surprise reelection in 1948 as a side effect of this campaign.[47]

Political warfare campaigns can require both international and domestic components to achieve their objectives. The DOS developed and orchestrated a comprehensive information and education program for the American public. The result was the ability to initiate the nascent American containment strategy. The Truman administration thus set the stage for the forty-plus years of containment that followed. Succeeding administrations oscillated between asymmetric and symmetric approaches to political warfare until the Cold War's dramatic conclusion in November 1989.

Political Warfare During the Reagan Years
The Reagan administration's approach to political warfare is worthy of thorough examination. First, because it defies easy categorization as asymmetric or symmetric, and second because it was high-risk and aggressive. Unlike his predecessors, President Ronald Reagan was never interested in merely containing the Soviet Union. Rather, Reagan stated, "My idea of American policy towards the Soviet Union is simple. It is this: we win and they lose."[48]

Although not fighting a war, the Reagan administration significantly increased defense spending to force the Soviet Union to negotiate nuclear arms control.[49] The result was the largest military spending increase in American history.[50] In real terms, the 1988 defense appropriation was higher than during the costliest year of the Vietnam War.[51] Brands observes:

The 1980s saw the most comprehensive US political warfare campaign of the entire Cold War. This offensive actually began under the Carter administration, which returned, through its focus on human rights, to highlighting the moral differences between East and West in a way Moscow found quite alarming.[52]

More than vast, Reagan's political warfare campaign was multimodal, employing all instruments of power. In May 1982, President Ronald Reagan signed the first US National Security Strategy. Its mandate was comprehensive and authorized the use of propaganda, military alliances, security cooperation, and trade expansion to contain and roll back the Soviet Union and the spread of communism.[53] The following year, the Reagan administration published National Security Decision Directive 75, "Relations with the Soviet Union." It states:

> US policy must have an ideological thrust which affirms the superiority of US and Western values of individual dignity and freedom: a free press, free trade unions, free enterprise and political democracy over the repressive features of Soviet communism.[54]

The moral dimension of the administration's political warfare campaign was instrumental. Recognizing the power of legitimacy in the eyes of domestic populations, President Reagan worked with Pope John Paul II to support the Solidarity movement in Poland[55] and took every opportunity to portray the Cold War in terms of good and evil. The Soviet Union was cast in the role of "The Evil Empire."[56] Thus, Reagan harnessed the influence of the pope on Poland's overwhelmingly Catholic population. Reagan further employed the language of "good and evil" to bolster the legitimacy of democracy while denying that of communism. This had the added advantage of comporting perfectly with American exceptionalism.[57]

On the economic front, the Reagan administration used a combination of sanctions and proscriptions on technology transfers against the Soviet Union to curtail its military capabilities and those of its critical energy industry, thereby further straining the already cash-strapped Soviet economy.[58] President Reagan's decision to invest in the Strategic Defense Initiative (known as Star Wars),[59] forced the Soviet Union to keep pace, placing even greater pressure on an already strained economy.[60] According to Reagan's former National Security Advisor, William Clark, a significant piece of the administration's political warfare strategy was to force the Soviet Union "to bear the brunt of its economic shortcomings."[61]

Although the Reagan administration imposed punishing sanctions on the Soviet Union in the early 1980s, the policy began under President Jimmy Carter in response to the Soviet invasion of Afghanistan. In addition to a grain embargo, Carter's sanctions included restrictions on the sale of oil and gas technologies, which made it difficult for the Soviet energy industry to keep pace with global innovation.[62] The combination of sanctions and proscriptions on technology transfers had their intended effect. At the end of the Cold War, US per capita income was nine times that of the communist bloc.[63]

The Enduring Lessons
Libraries are filled with books debating the effectiveness of Reagan's political warfare campaign. Whether Reagan "won" or the Soviet Union finally collapsed under the weight of its own internal contradictions is impossible to determine categorically. Nonetheless, the breadth, depth, and apparent effectiveness of the Reagan administration's political warfare campaign against the Soviet Union are impossible to discount. Hal Brands and Toshi Yoshihara conclude that:

> This was a top-to-bottom political warfare campaign, led by the president, supported by the bureaucracy and intended to exploit an adversary's weakness for competitive gain....they also threw the Kremlin off balance, reversed its ideological momentum, compounded its troubles in ruling its own restive empire, and thereby contributed to a dramatic turnaround in the superpower competition.[64]

The Past as Potential Prologue 79

Although imperfect, the US did run a successful, decades-long political warfare campaign against the Soviet Union, integrating disparate lines of effort that largely transcended partisan politics. The campaign relied on multiple political and military alliances, along with an unwavering commitment to free trade and democratic governance. To undermine the Soviet Union, the US used a combination of propaganda, information, and covert operations to exploit the weaknesses of the communist regime.

Contemporary Arguments Regarding Political Warfare

Over the past decade, countless practitioners and academics have called for a renewed American political warfare apparatus capable of addressing current threats. These threats are succinctly captured in the 2022 NSS:

> The need for a strong and purposeful American role in the world has never been greater. The world is becoming more divided and unstable.... The basic laws and principles governing relations among nations, including the United Nations Charter.... Democracies and autocracies are engaged in a contest to show which system of governance can best deliver for their people and the world.[65]

Given these challenges, calls for a renewed political warfare capability seem reasonable. What follows is a brief synopsis of the major arguments,[66] highlighting the similarities and differences between today's capabilities and those the US maintained during the Cold War. There are five primary arguments favoring a renewed American political warfare capability:

1. conventional war is too costly,
2. adversaries are outcompeting the US using political warfare,
3. an American political warfare capability would facilitate an affordable and integrated grand strategy,
4. adversaries are vulnerable to political warfare campaigns,

5. cyber operations, while new, are a necessity.

Conventional War Against A Near-Peer Competitor Is Too Costly.
Since 9/11, the US has been consumed with countering extremist organizations. By recent estimates, military interventions in Iraq and Afghanistan have cost more than $9 trillion, and the results have been, at best, disappointing.[67] That said, these costs pale in comparison to the potential toll of war with a near-peer competitor.

The result has been a tacit return to the Cold War logic of conflict. Anthony H. Cordesman writes,

> the failure to deter a general war would likely end in the equivalent of massive damage and to the de facto defeat of all the powers engaged. China and Russia understand this, and their competition focuses on a civil basis or at lower levels of gray area and hybrid operations, which involve far less risk....[68]

Jones argues, "First, US policymakers need to be prepared for significant inter-state competition to occur at the unconventional level since the costs and risks of conventional and nuclear war may be prohibitively high."[69] Michael J. Mazarr affirms this logic, concluding that the resurgent interest in gray zone (political) warfare stems from "...the desire to avoid direct confrontation."[70] During the Cold War, the specter of nuclear war and mutually assured destruction made conventional conflict between major powers anathema. That logic is again extant, prompting calls to develop America's political warfare capabilities.

Adversaries Are Outcompeting the US Using Political Warfare
The second argument is that the US is already losing this competition because it lacks an effective way to compete. Charles T. Cleveland et al. observe:

The Past as Potential Prologue

> The United States has proven to be ill-equipped to address the use of non-conventional means by revisionist, revolutionary, and rogue powers. The United States is seemingly impotent against cyberattacks targeting US government agencies, Russia's influence campaigns designed to destabilize the United States, Russia's penetration of the US electricity grid, and North Korea's nuclear posturing.[71]

Moreover, cultural tools employed against the Soviets—most notably the media industry—no longer pack the same punch they did during the Cold War. Kerry K. Gershaneck argues China has co-opted Hollywood to its ends, stating that "Beijing routinely demands that Hollywood portray China in a strictly positive light" and "punishes studios and producers that don't."[72] This results in an overwhelmingly positive portrayal of China within the entertainment industry and, by extension, globally.

The Requirement for an Integrated Political Warfare Capability

American Cold War strategy was integrative, using all instruments of national power. Over the last two decades, American grand strategy has become over-reliant on the military, increasing cost and decreasing effectiveness. Proponents of political warfare argue it would restore much-needed balance within American grand strategy. Mazarr observes:

> The primary importance of political warfare today is in terms of integrated strategies bringing together information operations, development aid, regime support, and other nonviolent options to encourage specific political outcomes.[73]

In addition to integration, revitalizing the government bureaucracy and its ability to direct political warfare is also required.[74] This includes ensuring that departments have the legal authorities to carry out their missions. Other authors advocate the establishment of a national political warfare center for "studying, understanding, and developing whole-of-government concepts of action ... for responding to nonconventional threats."[75] There is widespread agreement that an integrated political warfare strategy and capability is essential, whether the term "political warfare" is used to describe it or not. Ultimately, it is the capability rather than the terminology that matters.

Political Warfare: A Return to the Terminological Quagmire
Despite the relevance of its approach, the term "political warfare" itself will have to be discarded. Jones, once an advocate for the term (in 2018), later ignored it completely in favor of "irregular warfare" despite its similar meaning.[76] "Political warfare" is no longer the preferred term within academic and policy circles. The term "authoritarian political warfare" has begun to appear among NATO allies, casting political warfare as a nefarious technique perpetrated only by America's adversaries.[77]

Another potential explanation for the lack of enthusiasm for the term is Hoffman's argument that there is "no warfare" in political warfare.[78] Cordesman cautions, "The fact remains that there is no way to precisely define the differences between such operations, and a focus on creating a taxonomy which assumes that such rules exist is counterproductive."[79] He further concludes that "when one examines the full range of actual Chinese and Russian competition with the US, it becomes clear that there is no practical way to separate civil, gray zone, and hybrid operations."[80] What the capability is called matters less—what is more important is that it is reinvigorated as quickly as possible.

American Adversaries are Highly Vulnerable to Political Warfare Campaigns.
As explained in chapter 2, a fundamental pillar of US strategic culture is a belief in American exceptionalism. America continues to view itself as a "city on the hill" with the eyes of the world watching.[81] From the outset of the Cold War, the US drew upon this moral foundation to differentiate itself from the Soviet Union and formulate the central message of American political warfare: that of freedom. The US should

reaffirm its moral strength on the global stage and refresh that message for a contemporary audience.

The current competition is framed in terms as ideological as the Cold War. Today's adversaries are just as vulnerable as the Soviets were to information campaigns that decry the oppression of their own citizens, as well as those of their adversaries. From the plight of the Uyghurs to the invasion of Ukraine and Russia's continuing war crimes there, the message will resonate. Brands and Yoshihara conclude:

> For the United States to hold its own in today's great power competitions, it must think and act offensively in the realm of political warfare: it must do to its rivals what they are seeking to do to it. Taking the offensive is vital to exploiting the principal vulnerability of those rivals—their corrupt, illegitimate, authoritarian political systems.[82]

To exploit the weaknesses of its adversaries, America need only lean into its belief in democratic systems. American actions "should be consistent with another source of strength, the values enshrined in the American Constitution."[83]

Cyber Operations are Essential Political Warfare Tools
As much as the current era "rhymes" with the Cold War, there are substantial differences. One of the most consequential is the rise of cyberspace and cyber operations as a domain and key component of warfare, political and otherwise. Brands notes,

> It is generally less difficult and costly to conduct cyberattacks than to properly attribute and defend against them; it can also be cheaper and easier for a sophisticated adversary to spread disinformation than it is to detect and correct it.[84]

The centrality of offensive and defensive cyber capabilities, in conjunction with hardening vulnerability to cyber threats to today's information operations, is a key component of political warfare. Without it, America will remain critically disadvantaged. Jones concurs with this sentiment, adding a plea for the proper authorities to execute cyber operations.[85] Brands opines that "it will take another national mobilization" to build the cyber capabilities the US requires to compete against its adversaries.[86]

CONCLUSION

Although the circumstances the US and its allies face are not an exact match to those of the Cold War, there are enough similarities to use the principles of political warfare as an underpinning for American grand strategy. The task has never been more vital or more difficult to accomplish. The US must not ignore the domestic aspects of contemporary strategic competition. If it cedes this ground, it will lose. The term "political warfare" may be out of favor in the West, but its techniques are alive and well among American adversaries. Regardless of the name, the US must reinvigorate this capability as quickly as possible to remain competitive.

Notes

1. Although often attributed to Mark Twain, there is no evidence he ever said this. The earliest reference to this saying comes from Theodor Reik, "Essay 3: The Unreachables: The Repetition Compulsion in Jewish History," in *Curiosities of the Self,* 133.
2. White House, *[2022 National Security Strategy,* 6.
3. Ashford and Cooper, "Assumption Testing."
4. Munoz, "Toward Trade Opening," 65.
5. Becker, *US Soviet Trade in the 1980s,* 1; and Leonard, "China Is Ready," 127.
6. Shalal, "US Wants to End Dependence."
7. Martin, "China Relies on US."
8. Sitaraman, "A Grand Strategy of Resilience."
9. Becker, *US Soviet Trade in the 1980's,* 1.
10. ERP refers to the European Recovery Plan, more commonly known as the Marshall Plan.
11. Kennan, "Policy Planning Staff Memorandum." This chapter considers "political warfare," only in the traditional sense, meaning actions taken by states internationally to achieve their goals in the international arena. Increasingly, in the United States, the term is being applied to the partisan infighting that currently characterizes American politics. This is a separate development and is not considered in this chapter. For an example of this usage, see Naftali, "Political Warfare Comes Home." The origin of the term "political warfare" is often attributed to Lenin. However, a review of his writings and speeches does not yield an introduction or definition of the term. In the English translation of a November 1920 speech by Lenin, the term "political warfare," appears but is undefined. https://www.marxists.org/archive/lenin/works/1920/nov/21.htm It appears that Kennan attributed the term to Lenin from that speech. In the years following the Russian Civil War, one finds increasing references to Lenin's "school of political warfare"; see https://www.marxists.org/archive/lenin/works/1920/nov/21.htm.
12. Jones, *Three Dangerous Men,* 11.
13. "National Security Act of 1947."
14. Lucas and Mistry, "Illusions of Coherence."
15. Mahnken, "Containment," 133.

16. Brands, "The Dark Art of Political Warfare."
17. Hoffman, "On Not-So-New Warfare."
18. Lucas and Mistry, "Illusions of Coherence," 43.
19. Bamford, "The Points Were All."
20. Middaugh, "Homeward Bound."
21. Bamford, "The Points Were All."
22. Stewart, *American Military History*, 205.
23. Stewart.
24. Stewart, 227.
25. Pronk, "The Return of Political Warfare."
26. Brands and Yoshihara, "Waging Political Warfare," 21.
27. Boot and Doran, "Political Warfare."
28. Brands and Yoshihara, "Waging Political Warfare," 21.
29. Brands, "The Dark Art of Political Warfare," 4.
30. Lucas and Mistry, "Illusions of Coherence," 42.
31. Gaddis, *Strategies of Containment*, 27.
32. Worley, *Orchestrating the Instruments of Power*, 109.
33. National Archives, "The Truman Doctrine."
34. National Archives.
35. Gaddis, "Was the Truman Doctrine a Real Turning Point?" 389.
36. Worley, *Orchestrating the Instruments of Power*, 115.
37. Gaddis, "Was the Truman Doctrine a Real Turning Point?" 389.
38. North Atlantic Treaty Organization, "The North Atlantic Treaty."
39. Kennan, "Policy Planning Staff Memorandum," 669.
40. Brands, *The Twilight Struggle*, 25.
41. Miscamble, *George F. Kennan*, 72.
42. Hitchens, "Influences on the Congressional Decision," 51.
43. Hitchens, 52.
44. Hitchens, "Influences on the Congressional Decision," 53.
45. Hitchens, 53–55.
46. Hitchens, 52.
47. Hitchens, 56.
48. Mahnken, "Containment: Myth and Metaphor," 136.
49. Fischer, "Building Up," 165.
50. Fischer.
51. Zakaria, "The Reagan Strategy," 380.
52. Brands, "The Dark Art of Political Warfare," 21.
53. White House, "National Security Decision Directive 32," 1982, 1–4.
54. White House, "National Security Decision Directive 75," 3.

The Past as Potential Prologue 87

55. Worley, *Orchestrating the Instruments of Power*, 173.
56. Reagan, "Evil Empire Speech."
57. Heike, *The Myths That Made America*, 152.
58. Gordon, "The Law," 356.
59. Reagan, "The SDI Speech."
60. Gordon, "The Law," 536.
61. Gaddis, *Strategies of Containment*, 537.
62. Painter, "Energy and the End," 45.
63. Brands, *The Twilight Struggle*, 46.
64. Brands and Yoshihara, "Waging Political Warfare," 22.
65. White House, "The National Security Strategy," 7.
66. This section is the result of a comparative analysis of the arguments and conclusions presented in the following works: Brands and Yoshihara, "Waging Political Warfare;" Jones, "The Return of Political Warfare;" Brands, "The Dark Art of Political Warfare;" Cleveland et al., "The American Way of Political Warfare;" Gershaneck, *Political Warfare*; Boot and Doran, "Political Warfare;" and Cordesman, "The Broader Structure of US Strategic Competition with China and Russia; and Mazarr, *Mastering the Gray Zone*. Although this list is not exhaustive, it provides a representative example of current thought on developing an American political warfare capability among respected scholars and practitioners.
67. "Costs of the 20-year war on terror: $8 trillion and 900,000 deaths."
68. Cordesman and Hwang, *US Competition*, 11.
69. Jones, "The Return of Political Warfare," 32.
70. Mazarr, *Mastering the Gray Zone*, 50.
71. Cleveland et al., "An American Way of Political Warfare," 2.
72. Gershaneck, *Political Warfare*, 19.
73. Mazarr, *Mastering the Gray Zone*, 50.
74. Brands and Yoshihara, "Waging Political Warfare," 16.
75. Cleveland et al., "An American Way of Political Warfare," 1.
76. Jones, *The Return of Political Warfare*.
77. McInnis and Starling, *The Case for a Comprehensive Approach 2.0*, 4.
78. Hoffman, "On Not-So-New Warfare."
79. Cordesman and Hwang, *US Competition with China and Russia*, 12.
80. Cordesman and Hwang, 18.
81. Heike, *The Myths That Made America*, 152.
82. Brands and Yoshihara, "Waging Political Warfare," 16.
83. Cleveland et al., *Military Strategy in the 21st Century*, xvii.
84. Brands and Yoshihara, "Waging Political Warfare," 16.

85. Jones, *The Return of Political Warfare*, 32.
86. Brands, *The Twilight Struggle*, 249.

CHAPTER 5

TOOLS OF MILITARY STATECRAFT

> Military strategy, whether we like it or not, has become the diplomacy of violence.[1]
> Thomas Schelling

INTRODUCTION

Military force has always been central to geopolitics. Throughout history, war has served as a consistent means of tectonic political and social change—and thus, military scholars from Kautilya to Clausewitz have sought to chronicle the awesome and terrifying results of organized violence on a massive scale. Yet in a world where states are constantly on the brink of inflicting profound destruction upon each other, perhaps the most remarkable phenomenon is the non-use of military force— or, interestingly, the peacetime or non-kinetic use of military tools and personnel. When violence is held in reserve, states can use their militaries to threaten, deter, assure, posture, signal, and bluff, ultimately shaping the international environment to align with their national interests and political aims. This yields a profound but strange truth: military force is often most effective when it is not deployed at all.[2] Such is the paradox of modern military statecraft—the same toolset that is developed to be capable of unleashing dramatic upheaval in times of war can also be wielded usefully to support calculated maneuvers in times of what Americans would consider peace.

States have not always understood the non-kinetic utility of military statecraft. Since the advent of the Cold War, American policymakers have primarily oriented the military toward preparing for and conducting large-scale conventional wars. As mentioned in chapter 2, American strategic culture favors a distinction between war and peace. In the nineteenth and early twentieth centuries, the US military was primarily focused on North America—opening the frontier and maintaining the sanctity of the Monroe Doctrine—rather than engaging in expeditionary warfare.

However, the world wars served as a major inflection point in US military statecraft. American dependence on European defense industries in World War I prompted the development of a more robust mobilization capability and domestic defense industry. During World War II, newly devised financing arrangements under the Lend-Lease Act equipped and sustained Allied forces.[3] And for the first time, the US was at the center of coalition warfare, demanding the coordination of complex plans and activities across military services and nations.[4] Defense-industrial capacity, joint planning procedures, and the cultivation of interpersonal and organizational relationships among allies were essential to victory. This era also cemented the importance of overwhelming force as a core tenet of American military culture in the minds of both the military services and the bureaucracy.[5]

After World War II, the US sought to remold the international system—and provided the security backbone necessary for sustained peace. Thus, the Cold War saw expansion and increasing sophistication in the use of peacetime military statecraft. The maintenance of a significantly larger standing force, in tandem with new concepts like nuclear deterrence, forward defense, and tripwire forces, provided means of demonstrating

Tools of Military Statecraft 91

US commitment and achieving political objectives without resorting to war.[6] The period also saw greater reliance on international agreements and arms sales to build partner military capacity.[7] Not only was this diverse range of US military statecraft instrumental in ending the Cold War, but its importance expanded after the 9/11 terrorist attacks, as the DOD engaged in military operations, intelligence sharing, and capacity building with partner nations to combat extremist groups. Furthermore, the US military has frequently supported recovery efforts after natural disasters and humanitarian crises, both domestically and abroad.

The US has thus revolutionized military statecraft, forging a variety of tools to compete with adversaries below the threshold of armed conflict. It is thus a great irony that American policymakers are neglecting this impressive capability today. The majority of the DOD's $842 billion FY2024 budget is devoted to preparing for large-scale conventional and nuclear war.[8] Meanwhile, China and Russia are using asymmetric military means to offset and circumvent American advantages, including anti-access/area denial (A2/AD) capabilities as well as paramilitary forces designed to gradually change the status quo in their respective regions. These threats cannot be deterred or denied with conventional and nuclear forces alone; IW approaches are needed.

In an era of strategic competition and crises, increased use of the full gamut of military tools across the spectrum of conflict—from humanitarian operations through major power war—will increase American power, influence, and legitimacy far more than focusing on those tools which are primarily oriented toward winning major conventional wars. This chapter explores these tools by examining literature on military statecraft; relating such statecraft to power, influence, and legitimacy; presenting a snapshot of contemporary American military tools; and providing recommendations to improve its use.

THEORIES AND TOOLS OF MILITARY STATECRAFT

Defining military statecraft is relatively simple; to use Kyle J. Wolfley's definition, this would be "a state's use of military means to achieve foreign policy ends."[9] The term "military means" can be ambiguous since the lines between types of forces (i.e., military, paramilitary, and domestic police forces) can become blurred. In this chapter, the term "military means" embody the full set of tools, capabilities, and resources for leveraging organized violence against external adversaries—which in the US are placed under the purview of the DOD.

Unfortunately, the US lacks a comprehensive concept of military statecraft, and instead, its means are codified and classified as different programs and sub-elements in US legal code. Even within the NDS, many important tools like Key Leader Engagements are neglected or underappreciated,[10] and the DOD's siloed bureaucracy often prevents the coordination and integration of these means. This is especially problematic today. The most important role of the military is to fight (and prepare to fight) the nation's wars,[11] and the coordination of non-kinetic military tools can pose strategic dilemmas for US adversaries through coordinated peacetime IW campaigns. It is thus important to examine military statecraft holistically and understand the oft-neglected role of military statecraft in peacetime competition, which is the purpose of this chapter.

Given that militaries are oriented toward war but postured and utilized in peace, the theoretical relationship between war and peace is a central factor in exploring military statecraft. Therefore, we must first explore the questions: What is war? And how does it relate to the peacetime competition between states? One school of thought, as proposed by thinkers such as Sun Tzu and Kautilya, has long asserted that war and peace are interconnected and interpenetrating; thus, both kinetic and non-kinetic uses of military force are essential to military statecraft.[12] During the Cold War, Schelling built upon these theories by arguing that violence does not need to be applied against an enemy to be effective.[13]

Tools of Military Statecraft

Rather, violence can be demonstrated through coercive diplomacy, such as deterrence or compellence, and held in reserve to persuade the adversary to change or refrain from certain behaviors.[14]

So, peacetime military statecraft—including foreign basing, arms deals, exercises, and other means of signaling commitment to allies while deterring adversarial aggression—becomes more important than kinetic means in the conduct of competition short of war. As a result, the logic of military strategy has broadened beyond merely the realm of armed conflict, precipitating the dawn of peacetime grand strategy as a core facet of strategic competition.[15] Therefore, military statecraft is equally applicable in war and peace.

Another school of thought—more familiar to American policymakers—maintains that war and peace are distinct. Indeed, the beginning of a conventional war and its introduction of organized, kinetic violence unleashes irrational forces. These can be tempered by well-crafted policy, but only with great difficulty and uncertain outcomes. Thus, the careful manipulation of kinetic and non-kinetic means of military statecraft, often exercised in peacetime, becomes substantially more difficult to implement and signal during war.

Moreover, American strategic culture adheres to the belief that war is not only distinct from peace but that, normatively, it should be distinct from peace. To paraphrase Rosa Brooks, without boundaries between war and peace, everything will become war, and the military will become everything.[16] Domestic laws and norms can weaken, and society may rely too much on its military for the execution of its foreign policy.

To succeed in an era of crises and strategic competition, a more nuanced approach is required. While the peacetime and wartime uses of military statecraft are very different, as argued by the second school of thought, they are equally essential to the conduct of IW due to the interconnected relationship between war and peace. Such tools then can and should be used across wartime and peacetime scenarios. War and peace exist on a continuum, and placing black-and-white distinctions between them puts the US at a significant disadvantage in an era of strategic competition. In fact, the ability to threaten violence is vital during peacetime competition and often shapes the decisions of friends and foes alike.

Non-kinetic military means of demonstrating commitment, deterring adversaries, and assuring allies are, therefore, essential to achieving objectives like power, influence, and legitimacy below the threshold of armed conflict. Moreover, such non-kinetic tools embody military statecraft at its most rational[17] since signals and messages can be calculated carefully in peacetime while accompanied by non-military means. Unfortunately, however, the US often prioritizes wartime applications of military force over peacetime applications. Given that modern military forces spend so little time fighting, and that their jurisdiction is far greater than the prosecution of conventional operations, this chapter focuses on non-kinetic, peacetime roles for military statecraft within broader IW strategies.

The literature on military statecraft provides alternatives for categorizing military tools, including Wolfley's distinction between traditional (kinetic) and shaping (non-kinetic) uses of military force[18] or typology by purpose, as delineated by Melanie W. Sisson, James A. Siebens, and Barry M. Blechman.[19] However, the former is too simplistic and obscures the nuance of the US diverse military toolset, while the latter is undermined by the reality that military tools can simultaneously have different and overlapping purposes. Therefore, this chapter suggests an alternative typology based on the level of interaction at which the military aims to achieve any given objective (figure 5).

Tools of Military Statecraft

Figure 5. Typology of Military Statecraft Tools.

Interpersonal Level

Tools that cultivate person-to-person ties between allied, partner, and (occasionally) adversary military personnel, building greater affinity and understanding

Sample Tools:
- Key Leader Engagements
- Friendly visits
- Military personnel exchange programs
- International Professional Military Education, including:
  - International Military Education and Training

Contribution to Irregular Warfare goals:
1. Power: Increases the competence of allied junior officers
2. Influence: Cultivates greater cultural affinity among militaries
3. Legitimacy: Improves perceptions of U.S. military officers

Organizational Level

Tools that enable military institutions to support each other, thereby synchronizing doctrine and procedures, building interoperability, and demonstrating mutual commitments

Sample Tools:
- Security cooperation, including:
  - Foreign Military Sales
  - Foreign Military Financing
  - Technological cooperation
  - State Partnership Program
- Exercises
- Humanitarian operations
- Peacekeeping operations

Contribution to Irregular Warfare goals:
1. Power: Empowers partners to degrade adversary capabilities (i.e., bloodletting)
2. Influence: Reassures allies and partners by demonstrating robust commitments
3. Legitimacy: Supports states in crisis through peacekeeping and humanitarian operations

Systemic Level

Tools that allow states to guarantee their security, deter their adversaries, and, if necessary, fight major wars—in accordance with more traditional uses and threats of violence

Sample Tools:
- Conventional forces
- Nuclear forces
- Special operations forces
- Global force posture, including:
  - Basing
  - Rotational deployments
- Technological innovation
- Acquisition and budgetary reform

Contribution to Irregular Warfare goals:
1. Power: Builds the unilateral conventional and nuclear power of a state
2. Influence: Deters adversaries by shaping their perceptions about the utility of force to achieve revisionist ends
3. Legitimacy: Enables states to support international law with military force

Military Statecraft and Power, Influence, and Legitimacy

If military statecraft is to serve a broader IW strategy, it must be calibrated to increase the power, influence, and legitimacy of the US and its allies and partners or to decrease those of its adversaries.

Military Power

Military power is often the primary objective of military statecraft as it enables the achievement of a variety of national interests and political aims. According to Barry R. Posen, military power is not an abstract concept nor a definitive absolute value.[20] Even the most powerful military in the world cannot pursue every mission and achieve every objective. Instead, military power is more nuanced and has three main characteristics: it is dynamic in that it is "susceptible to constant, and sometimes dramatic, change;" it is situational in that it varies "not only with the given time period but also with the particularities of situation and geography;" and most importantly, it is defined relative "to other states' capacities to employ military means directed to the same or related objectives."[21] Hence, military power is increased by building competitive advantages over adversaries, maintaining a vast array of capabilities at every level of escalation, and diversifying military statecraft to achieve different political objectives. Such power allows states to effectively shape adversaries' behavior, including deterring strategic attacks, compelling adversaries to cease revisionist behavior, or achieving objectives through the application of military force.

A state can possess both latent and active sources of military power. Kenneth Waltz suggests that latent determinants include "size of population, territory, resource endowment, economic capability, military strength, political stability, and competence."[22] But there are also a variety of tools with which policymakers can actively build military power. These include conventional and nuclear forces designed to deter adversaries: by denial, preventing them from achieving their goals; by cost imposition, threatening to punish aggressive behavior; by resilience, withstanding adversary attack; and by escalation dominance, dissuading adversary escalation in war due to superior capabilities.[23] In an era of crises and strategic competition, resilience, in particular, is increasingly important, and would include initiatives like hardening military equipment and facilities against the impacts of increasing severe weather and cyberattacks, as well as training service members to recognize when they are being targeted by adversarial disinformation campaigns. Beyond deterrence, many of the aforementioned capabilities are also useful for compelling changes to adversary behavior and, if necessary, for warfighting.

While not the primary focus of this book, conventional and nuclear forces merit attention as a backstop to all IW activities. If the US pivots away from these capabilities, adversaries will return to more overt forms of aggression like Russia's invasion of Ukraine. Thus, the US must continue to support, modernize, and posture its conventional and nuclear forces to deter adversary activities above the threshold for war while also waging IW campaigns below the threshold of war. At the same time, merely possessing such forces is insufficient to deter competitor IW activities.

In IW, there are other military tools that contribute to greater relative power. This includes Mearsheimer's concept of "bloodletting"[24] as a means "by which a country increases the costs to its rival by making a conflict in which that rival is involved more difficult."[25] Through either covert support for opposition groups, overt arms sales, or broader security cooperation vis-à-vis countries at war, bloodletting embodies a low-cost means to degrade an adversary's military capabilities without directly employing its own armed forces, thereby transforming an adversary's struggle into a proxy conflict.[26] Another tool for expanding military power is building upon multilateral and bilateral alliances to increase the collective military capabilities of like-minded nations. Finally, arms control agreements embody an opportunity to codify a favorable status quo (or to forestall an unfavorable status quo) and mitigate arms race pressures.

Military Influence

The 2022 NDS's concept of campaigning captures the essence of building military influence, which are activities that "improve our baseline understanding of the operating environment and seek to shape perceptions" of friends and foes alike.[27] This includes extending deterrence to allies "by sowing doubt in our competitors,"[28] but more interestingly—and in contradistinction to military power—it prioritizes the reassurance of allies and partners. By demonstrating robust security commitments and altering perceptions, a state can shape and restrain the behavior of its allies.[29] For example, by extending its nuclear umbrella to states like Germany and Japan, the US has contributed to a perception of greater security among these countries, thereby dissuading them from building their own nuclear arsenals and sparking regional arms races.[30]

American military influence is advanced by a variety of tools, including security cooperation to ensure that allies and partners buy US weapons systems and depend on US assistance, exercises to influence allies' perceptions of coalition warfare and advance interoperability, information sharing and intelligence cooperation to build a common operating picture, and training and education programs that foster the exchange of ideas and techniques. Overall, "the primary value of security assistance programs is in their cultivation of relationships, the building, maintenance, and management of which is hoped to go some distance in inducing cooperation where and when the US would like it."[31] It is important to note, however, that just as the US can use these tools to shape allied beliefs and behavior, allies can also leverage such US commitments to restrain American behavior.[32] Nonetheless, the benefits far exceed the costs, as America's alliance system is one of its greatest competitive advantages.

Military Legitimacy

The essence of US military legitimacy is convincing its domestic constituency as well as the international community that the US uses its military responsibly and in accordance with international laws, values, and ethical standards.[33] Another element of US legitimacy is the ability to respond appropriately during crises. The US military, for example, may be asked to respond to weather-related disasters with hospital ships overseas. The National Guard (NG) may be required to provide support during domestic hurricanes or wildfires, or even terrorist attacks. If those resources are deployed quickly and efficiently, they bolster legitimacy.

Since legitimacy is oriented toward perceptions, both actions seen as morally "right," such as acts of goodwill (e.g., international law enforcement), or "wrong," such as the torture of combatants, reinforce or diminish legitimacy, respectively.[34] Thus, freedom of navigation operations (FONOPs) in defense of the global commons, humanitarian and peace operations, counterdrug assistance, military information support operations (MISO) designed to counter radicalization, and a variety of friendly visits and Key Leader Engagements (KLEs) all bolster military legitimacy. Conversely, improper use of the military instrument not only degrades domestic perceptions of legitimacy but those internationally as well. For example, the documented war crimes perpetrated by the Russian military in Ukraine have further delegitimized an already suspect force.[35]

A SNAPSHOT OF PROMINENT MILITARY TOOLS

This section assesses how the US has used and misused military statecraft to achieve power, influence, and legitimacy, dividing these tools along the aforementioned typology: interpersonal, organizational, and systemic.

Interpersonal Tools

Interpersonal tools are often neglected in the literature because they do little to advance US power writ large, and their impact is hard to quantify. Nonetheless, they are valuable means of increasing influence and legitimacy among allies and partners. Only individual engagements can contribute to cultural affinity and familiarization among different countries' military forces (a requirement discussed in chapter 4). Such tools include KLEs, which occur when high-ranking civilian or uniformed personnel visit a foreign country, embodying "a special form of diplomacy that carr[ies] implicit messages about the security relationship between the US and the host nation, and more generally about the extent of US interest in the host nation's region."[36] Similar effects can be achieved through friendly visits of US forces, including port calls and participation in allied and partner airshows.[37]

The International Professional Military Education (IPME) enterprise, including the International Military Education and Training (IMET) and other programs, provides a structured vehicle to impart expertise, facilitate joint problem-solving, develop human capital, and build military influence among allies and partners.[38] These programs are considered the gold standard internationally, attracting the best and brightest talent across the globe.[39] For those students who spend time in the US, the influence-building component cannot be underestimated. Furthermore, the relationships built at this interpersonal level have strategic impacts for decades to come. Annually, the IMET trains approximately 78,000 individuals,[40] who come from more than 100 countries,[41] demonstrating the span of US interpersonal military tools.

An underappreciated indicator of military strength is the quality of junior leaders (officers with less than ten years of experience) and noncommissioned officers (NCOs)—the individuals often tasked with executing complex battlefield tactics and operations. In war, their ingenuity, resourcefulness, and leadership can mean the difference between victory and defeat. America's ability to trust its junior leaders, particularly NCOs, following decentralized mission command, is a significant advantage over the rigid and hierarchical military structures of authoritarian adversaries like China and Russia.[42]

Tools of Military Statecraft   101

Thus, training the junior leaders of other militaries can help impart this US advantage to allies while also learning from their best practices. Accordingly, programs like the NATO Military Personnel Exchange Program assign individual officers to leadership positions within another country's military for a specific period.[43] Such exchanges may become semi-permanent; for example, the US–South Korean agreement that places Korean soldiers in US formations (KATUSAs) has endured for seventy years.[44]

Organizational Tools
Alliances are an effective means of increasing military power and influence, and America's global network of partners and allies remains a key competitive advantage. However, the warfighting capabilities of alliances are only one component of a vast and important architecture. Most of the tools necessary for sustaining alliances exist at the organizational level, primarily in the form of security cooperation.

Security cooperation is an umbrella term referring to the many ways in which the US "assist[s] partners in growing and enhancing their own capabilities."[45] It is a core means of enhancing both the military power of alliances and American influence with those alliances. It is also vital for prevailing in an era of crises and competition.

There are a variety of security cooperation tools available to policymakers, all of which are oriented toward assuring allies and partners, building their military capacity, enhancing interoperability, and influencing their behavior. Most prominent is Foreign Military Sales (FMS), or exports of military equipment and corresponding services like training and maintenance.[46] In 2022 alone, US FMS totaled $50.1 billion.[47]

The US also provides military aid in the form of financial assistance, loans, or grants through Foreign Military Financing (FMF). While Russia and China can often compete with American FMS by selling their weapons systems, American FMF is unparalleled. For example, between 2013 and

2018, the US provided $35 billion to 71 countries, while China only donated $560 million to 35 countries.[48]

Other security cooperation tools include "advise and assist" missions to provide combat training, intelligence, and logistical support;[49] "train and equip" missions to build partner capacity in pursuit of counterterrorism, counterdrug, and other activities;[50] and defense institution building (DIB) to strengthen partners' defense ministries through institutional reform.[51] DIB is vital to institutionalizing the training and education that happens at the interpersonal level. This includes helping partners to build their own training capacity, advising on policies such as promotion criteria, and sharing standard operating procedures that devolve authority to lower levels as appropriate.

More recently, alliances and partnerships have advanced technological cooperation, such as the longstanding National Technology Industrial Base[52] and the Australia-UK-US (AUKUS) agreement.[53] Interoperable systems are vital for working with allies and partners across the spectrum of conflict. These kinds of interorganizational relationships facilitate interoperability.

There is also the State Partnership Program (SPP), which partners US state-NG units with foreign militaries to "facilitat[e] cooperation across all aspects of international civil-military affairs and encourag[e] people-to-people ties at the state level."[54] This has been essential to building Ukrainian capacity. Indeed, the 1993 pairing of the California NG with Ukrainian military forces cemented a deep level of familiarity. After Russia's 2014 annexation of Crimea, this existing relationship provided the foundation for a rapid expansion in military exercises, a growth in organizational and interpersonal connections, and the refinement of military doctrine. The SPP is a muscle that can be activated at any time and with great flexibility—while the relationship began with California, other NG units have rotated through Ukraine over time. The SPP helped prepare Ukrainian military forces for conventional war with Russia.[55]

Tools of Military Statecraft 103

Furthermore, the SPP is not constrained solely to conventional military capabilities. Specific partnerships also focus on disaster and emergency response, cyber defense and communication security, counterterrorism, and building medical competencies. In fact, the training and development of these capabilities have paid dividends even inside the US. During the COVID-19 pandemic, a military medical team from Poland deployed to Chicago to assist healthcare workers as part of the Poland-Illinois SPP.[56]

In addition to arming and training allies and partners, organizational military tools include the actual practice and conduct of combined operations, primarily through military exercises.[57] US and allied exercises were originally oriented toward preparing for conventional conflict, but they now serve broader political objectives including attraction (to draw other states into closer military relationships), socialization (to influence the roles, norms, and practices that partner militaries will adopt), delegation (to encourage greater burden-sharing), and assurance (to reduce insecurity and shape perceptions and behavior of partner militaries).[58] When combined with information-sharing and intelligence cooperation, exercises can bolster US influence among allies.

Other tools at this level are designed to influence the geopolitical environment and increase US legitimacy among friendly militaries and local populations. For example, stability operations provide security to the local populace in war-torn regions or fragile states.[59] US military force can also enforce international law, either unilaterally through FONOPs, which garner goodwill with vulnerable states,[60] or multilaterally through UN Peace Operations.[61] Finally, US military legitimacy is bolstered by tools such as counterdrug assistance, MISO to counter terrorist radicalization and recruitment efforts, and Overseas Humanitarian, Disaster Assistance, and Civic Aid (OHDACA) operations to support populations in crisis.

Systemic Tools
Systemic tools designed to increase military power lie at the heart of US military statecraft. Often kinetic in nature, these tools are the

focus of major DOD missions and budgetary allocations. Conventional forces are a primary tool for expanding military power at the systemic level. Through technological, doctrinal, and organizational innovation in the 1970s and 1980s, the US military advanced substantial operational innovations, drawing upon global surveillance, precision strike, and stealth capabilities.[62] Competitors were left stunned by their success during the Gulf War. But over the last thirty years, adversaries have learned from this experience[63] and have developed A2/AD capabilities like ballistic and cruise missiles to prevent the US from accumulating combat power and concentrating its forces in distant regions.[64] To maintain credibility in its power-projection capabilities and extended deterrence commitments, the US must continually modernize by adopting transformative dual-use technologies while simultaneously developing new ways of fighting. This requires acquisition and budgetary reform,[65] or at least changes to existing incentive structures,[66] to streamline innovation adoption processes.

Nuclear weapons are another systemic tool designed to project power and deter strategic attacks against the US and its allies. America's longstanding nuclear advantage[67] is now under threat, raising questions about the necessary quantity and quality of its arsenal.[68] In the wake of Chinese nuclear expansion, "for the first time, the United States will simultaneously contend with two major nuclear powers,"[69] stretching the limits of its counterforce capability. Additionally, Russia's exotic nuclear weapons[70] threaten to provide a coercive advantage at lower rungs of the escalation ladder, provoking debates about the necessity of US tactical and/or low-yield nuclear weapons to regain escalation dominance.[71] Despite these challenges, nuclear deterrence has been a successful tool of military statecraft, as these devastating weapons have not been used since 1945.

However, the recent collapse of arms control regimes like the Intermediate-Range Nuclear Forces Treaty and the growing nuclear capabilities of multiple states suggest that another nuclear arms race is possible.

Furthermore, the advent of new capabilities like hypersonic weapons, directed energy, and sophisticated air and missile defenses may undermine the nuclear triad (the strategic combination of air-, sea-, and ground-based nuclear weapons), requiring a broader conception of the emerging strategic forces balance.[72] The continued push to acquire nuclear weapons by Iran must also be considered in this calculus.

Another important systemic tool is global force posture (the strategic distribution and deployment of military assets), as well as the basing and rotational deployments necessary for sustaining it. Force posture is an essential means of signaling commitment, deterring adversarial aggression, and reassuring allies and partners. The use of tripwire forces is particularly "significant not only because it projects military capabilities in case they are needed on short notice, but also because it means that if that country were to be invaded, the US would be involved in the conflict automatically."[73]

The post–Cold War shift from permanent to rotational deployments has considerably decreased expenditures but diluted US military presence abroad, and hence its respective influence. Adversaries have exploited this trend by using paramilitary forces to secure greater regional influence. For example, the PAFMM harasses local fishermen to enforce illegal claims in the South China Sea, while Russian mercenaries like the Wagner Group operate in Europe, Africa, and the Middle East.[74] The US needs greater access to deter territorial revisionism while taking steps to increase the resilience and survivability of forward-deployed assets.[75]

Recommendations

The following recommendations address improving US military statecraft as a means of IW with the objectives of gaining relative power, influence, and legitimacy.

Employ All Military Statecraft Means
Policymakers are too often focused solely on kinetic military tools. Deterring and preparing for large-scale conventional war should remain an important priority, but greater attention must be paid to the gamut of other options. Due to a long history of experimentation, investment, and innovation in the Cold War, US military statecraft is broader and deeper than that of China and Russia. Rather than forfeit this advantage, policymakers should make it a central facet of US strategy. This necessitates a more holistic approach to military statecraft—accordingly, policymakers should define military statecraft in key documents like the NDS, while more clearly integrating the available kinetic and non-kinetic tools to foster widespread familiarity with their utility. The 2022 NDS, with its focus on Integrated Deterrence and Campaigning, is an admirable attempt to expand the available means but requires greater specificity to truly realize their impact.

Redefine Security Cooperation
As a term, security cooperation has become so bloated and overused that it has almost lost its meaning. Policymakers should instead simply use one umbrella term for all military tools: military statecraft. That accomplished, the DOD should separate and categorize the key activities and tools formerly grouped under security cooperation to more easily select among priorities and tradeoffs in these lines of efforts. These tools can be tiered based on the nature of the unique relationships, as follows:

- **Security Collaboration:** to describe the multilateral and bilateral relationships among the US and its most trusted allies. Tools to strengthen and maintain these alliances would include sharing intelligence, offering robust FMS and FMF, jointly developing high-end technologies, conducting major exercises, and participating in the development of war plans. This would also cover the significant reforms to the overly restrictive International Trade in Arms Regulations that currently prevent the most important security collaboration initiatives.

- **Security Cooperation**: to describe the bilateral relationships between the US and its partners (those who are not treated as allies and should not receive the most sensitive information from the US but merit closer cooperation). Tools to improve these relationships would include limited joint exercises, some FMS and FMF, military personnel exchanges, and training programs.
- **Security Coordination**: to describe the bilateral relationships between the US and its competitors. Tools to avoid conflict include KLEs, arms control agreements, and other means of coordination, while conventional and nuclear forces are postured to maintain deterrence.

Reprioritize Security Cooperation

Reframing security cooperation is an important first step, but reprioritization of the US finite resources is also required. To address existing shortfalls, there must be careful consideration of both which states receive military assistance and what is provided. Furthermore, analyzing and measuring the impacts of security cooperation should drive future US spending priorities.

FMS and FMF continue to demonstrate outdated priorities. Though the current military assistance to Ukraine dwarfs the US support to every other state, this is the first time a European country has held the top spot since the Truman administration.[76] For decades, the preponderance of US military aid has been funneled to Middle Eastern states like Israel, Egypt, and Jordan rather than to allies and partners in priority regions like the Indo-Pacific and Europe.[77]

Such security cooperation efforts are not always successful in influencing allies. In the lead up to the US invasion of Iraq in 2003, Turkey would not provide materiel support or basing.[78] More recently, Turkey decided to acquire the Russian-manufactured S-400 missile system in 2019, complicating NATO interoperability.[79] Security cooperation programming and the level of US assistance should be consistently reviewed in terms of prioritization to ensure that it is achieving its intended purpose.

Even when security cooperation successfully increases US military influence, it can still fail to produce greater military power. For example, FMS is often oriented toward sophisticated weapons systems such as the F-35 due to their status and prestige, but vulnerable states like Taiwan would benefit more from lower-cost, asymmetric missile and air-defense capabilities that would actually build resilience in the face a *fait accompli* attack.[80] The success of asymmetric capabilities like anti-tank, anti-air, and artillery systems in the Russo-Ukrainian War should be a wake-up call. Military influence should not be bought at the expense of relevant military power.

Unfortunately, building allied and partner military capacity has produced mixed results in recent years. Efforts to bolster Ukrainian forces —both before and after the large-scale Russian invasion in February 2022—have proven successful. But security cooperation with the Iraqi military in the wake of the 2003 US invasion extending for more than a decade produced only moderate results, while efforts in Afghanistan lasting nearly twenty years failed entirely.[81] These cases demonstrate that security cooperation can be more effective when it builds upon existing military institutions, when it facilitates more robust interpersonal relationships, and when it encourages a healthy balance between partner dependency and self-sufficiency.

Reform Acquisition Processes
Military tools at the organizational level are often impeded by overly bureaucratic processes and byzantine structures. The system is a jumble of acronyms, authorities, and overlapping jurisdictions—both between the DOS and the DOD, as well as among DOD offices. During the Cold War, organizational efficiency was not a concern—but in an era of strategic competition and crises, policymakers no longer have that luxury. These processes have delayed key aid to Ukraine. To compete with Chinese and Russian reforms, it is imperative for the DOD, in tandem with other interagency actors, to improve the bureaucratic processes and structures underpinning its military statecraft.

Spark Defense Innovation
Confronted by multiple threats above and below the threshold of war and with impressive but still limited resources, the US needs to offset the quantity and quality of its adversaries' conventional forces with a superior technological edge. This will free up resources for IW.

The DOD's innovation ecosystem is a key tool of competition but remains a Cold War relic. The innovations most likely to revolutionize the future of warfare will not originate from government labs but rather from the start-ups, scale-ups, and venture capital firms at the forefront of today's dual-use technologies. Therefore, to spark innovation and adopt new technologies that will increase military power at the systemic level, the DOD must help smaller companies bridge the so-called "valley of death" between new technologies and major defense programs. In particular, policymakers must: increase the flexibility of defense spending beyond the rigid Planning, Programming, Budgeting, and Execution process; alter the incentive structures inherent in the defense acquisition workforce and empower savvy program managers to navigate byzantine resourcing and acquisition cycles; and build upon the success of new organizational models of innovation, such as the Defense Innovation Unit and Space Development Agency, by providing them greater resources and authorities to scale up emerging technologies.[82]

Conclusion

Military statecraft represents a highly competitive arena and a lower-cost alternative to great-power war. While competitors have made significant inroads, America retains abundant advantages. US policymakers must build upon the interpersonal, organizational, and systemic tools of military statecraft to bolster America's influence and legitimacy, just as kinetic tools maintain US military power.

Notes

1. Schelling, *Arms and Influence.*
2. Schelling.
3. Herman, *Freedom's Forge.*
4. Matloff, "Allied Strategy in Europe, 1939–1945."
5. Strachan, *The Direction of War*, 19.
6. Schelling, *Arms and Influence.*
7. Defense Security Cooperation Agency, "Organization History."
8. US Department of Defense, "Department of Defense Releases."
9. Wolfley, "Military Statecraft and the Use," 3.
10. US Department of Defense, *2022 National Defense Strategy.*
11. Clausewitz, *On War.*
12. Sun Tzu, *The Art of War*; and Kautilya, *Kautilya's Arthashastra*, 517.
13. Schelling, *Arms and Influence*, 1–2.
14. Schelling.
15. Freedman, "Strategy."
16. Brooks, *How Everything Became War*, 342.
17. Heuser, "Clausewitz's Ideas."
18. Wolfley, "Military Statecraft and the Use," 4.
19. Sisson, Siebens, and Blechman, *Military Coercion*, 16–32.
20. Posen, "Command of the Commons."
21. Jordan et al., *American National Security*, 272.
22. Waltz, "Theory of International Politics," 131..
23. Schelling, *Arms and Influence*; US Department of Defense, *2022 National Defense Strategy.*
24. Mearsheimer, *The Tragedy.*
25. Stuster, "The Strategic Logic."
26. Christian Trotti, "The Return; Berman and Lake, *Proxy Wars*, 1-28].
27. US Department of Defense, *2022 National Defense Strategy*, 12.
28. US Department of Defense, *2022 National Defense Strategy*, 12.
29. Jordan et al., *American National Security*, 280.
30. Kroenig and Trotti, "Modernization as a Promoter."
31. Sisson, Siebens, and Blechman, *Military Coercion*, 23.
32. Jordan et al., *American National Security*, 280.
33. Jordan et al., 277.
34. Sisson, Siebens, and Blechman, *Military Coercion*, 22–27.

Tools of Military Statecraft

35. MacDonald, "Russian Abuses in Ukraine."
36. Sisso, Siebens, and Blechman, 25.
37. Sisso, Siebens, and Blechman.
38. Defense Security Cooperation Agency, "Chapter 10 - International Training."
39. Kurlantzick, "Reforming the US International Military Education."
40. Defense Security Cooperation Agency, "International Military Training."
41. Fabian, "US IMET Participation," 242.
42. Garamone, "Noncommissioned Officers."
43. Suits, "Building Relationships."
44. "KATUSA Soldier Program."
45. Sisson, Siebens, and Blechman, *Military Coercion*, 22.
46. Krause, "Military Statecraft."
47. Hasselbach, "SIPRI: US Arms Exports Skyrocket."
48. Beauchamp-Mustafaga, "China's Military Aid."
49. Sisson, Siebens, and Blechman, *Military Coercion*, 23.
50. Defense Security Cooperation Agency, "Global Train and Equip" and "Section 333 Authority to Build Capacity."
51. Perry et al., *Defense Institution Building*.
52. Greenwalt, *Leveraging the National Technology Industrial Base*.
53. US Department of Defense, *2022 National Defense Strategy*, 14–15.
54. National Guard, "State Partnership Program."
55. Greenhill, "State Partnership Program Turns 30."
56. US European Command Public Affairs, "USEUCOM State Partnership."
57. Sisson, Siebens, and Blechman, *Military Coercion*, 17–22.
58. Wolfley, "Military Statecraft and the Use."
59. Jordan et al., *American National Security*, 331.
60. Sisson, Siebens, and Blechman, *Military Coercion*, 27–28.
61. Jordan et al., *American National Security*, 332.
62. Van Atta et al, "Transformation and Transition."
63. Colby, "Testimony Before the Senate."
64. Montgomery, "Contested Primacy"; and Hammes, *An Affordable Defense*; US Department of Defense, *2022 National Defense Strategy*, 4.
65. Commission on PPBE Reform, *Commission on Planning*; Lofgren, McNamara, and Modigliani, *Atlantic Council Commission*.
66. Grundman, *The Monopsonist's Dilemma*; and Trotti, "To Centralize?"
67. Kroenig, *The Logic of American Nuclear Strategy*.
68. Peters, "America's Current Nuclear Arsenal."

69. Chairman of the Joint Chiefs of Staff, *2022 National Military Strategy*, 2.
70. Kroenig, Massa, and Trotti, "Russia's Exotic Nuclear Weapons."
71. Insinna, "Biden Administration Kills."
72. Pavel and Trotti, "New Tech Will Erode."
73. Sisson, Siebens, and Blechman, *Military Coercion*, 17.
74. Babbage, *Winning Without Fighting*, 39–40.
75. Crouch, Pavel, Starling, and Trotti, "A New Strategy."
76. Masters and Merrow, "How Much Aid?"
77. Lurye, "10 Countries."
78. "Turkey rejects US troop proposal."
79. Ciddi, "Is Turkey About to Ditch?"
80. Hammes, *An Affordable Defense*.
81. Byrd, "Why Have the Wars?"
82. Commission on PPBE Reform, *Commission on Planning*; Lofgren, McNamara, and Modigliani, *Atlantic Council Commission*; and Trotti, "To Centralize or Not to Centralize?"

Chapter 6

Tools of Economic Statecraft

> International economic exchange is one of the most spectacularly successful examples of international influence in history; yet is rarely so described.[1]
>
> David A. Baldwin

Introduction

Economic statecraft has always been a vital component of national security.[2] For millennia, great powers have risen and fallen based on their ability to accumulate wealth, manage resources, and wield economic leverage.[3] Yet, over the last century, the role of economics in national security has become far more complex. After World War II, America led the establishment of the Bretton Woods institutions, including the World Bank, the International Monetary Fund (IMF), the International Finance Corporation (IFC), and the General Agreement on Tariffs and Trade (GATT)—the precursor to the World Trade Organization (WTO). These institutions and those that have followed them—the Organization for Economic Co-operation and Development (OECD) and the Financial

Action Task Force (FATF)—have vastly expanded the economic toolset available to states, especially those at the global economy's center.

Because the architecture of this world order rests upon the global dependence on the US dollar, the United States has enjoyed outsized influence and power.[4] Economic institutions have enabled the US to employ both the proverbial carrot (through access to resources, loans, and institutional influence) and the stick (through sanctions and the threat of denying trade and finance).[5] Additionally, American support for capitalism and free market principles have lent legitimacy to US actions.

In the 1990s, optimism about the pacifying and enriching prospects of international economic institutions, free-flowing trade, and investment abounded.[6] There was a belief that increasing economic interconnection and deeper integration into the institutions underpinning the liberal, rules-based international order were mutually reinforcing trends, encouraging states to adopt more cooperative foreign policies.[7] However, over the last decade, the convergence of global economic interdependence, a rising wave of protectionism, and surging strategic competition have prompted states to employ economic tools as a means of power and influence instead of mutual wealth creation, thereby weakening the liberal international order.[8]

Given America's central role in the global economy and its adversaries' increasing economic savviness, instrumentalizing economic statecraft is essential to both strategic competition and prevailing in an era of crises. Thus, this chapter outlines the available economic tools and potential ways the US can better use the tools in pursuit of power, influence, and legitimacy. It then examines the current posture of American economic statecraft. Finally, it makes recommendations as to how the US should leverage economic tools to better compete with China and Russia and insulate itself from crises.

Economic Strength and Tools of Economic Statecraft

Like military statecraft, a state's ability to use economic tools depends upon the strength of its economy, which includes resources such as labor, technology, natural resources, capital (factories, infrastructure, etc.), reserves of international liquidity, and foreign investments.[9] These factors are the product of a state's natural endowments—history, geography, and culture, as well as domestic and foreign policies.[10] For example, potentially enticing domestic policies such as government seizure of land, labor, or capital can wreak severe international consequences—including decreased foreign direct investment (FDI) as other states grow wary of instability, dampening entrepreneurship and inhibiting growth.

Conversely, policies such as incentivizing investment in human capital and technological growth, expanding physical and digital infrastructure, and establishing legal regimes to define and enforce property rights—while avoiding excessive regulatory burdens that can inhibit innovation—all contribute to a more productive economy. Such policies help generate economic strength, attract FDI, and enable military power. In addition to pro-growth domestic policies and a stable domestic financial industry, economic growth depends upon savvy foreign policies—including trade agreements and integration into international economic institutions. For the last 70 years, economic interdependence has facilitated American prosperity as well as those of its closest allies.

A state can leverage its economic strength to achieve political, strategic, or security goals by using three types of economic tools:

- trade-based tools: the tools related to the manipulation of international trade relationships;
- capital-based tools: tools related to control over endowments of capital and financing; and
- domestic economic policies.

Figure 6. Typology of Economic Statecraft Tools.

Trade-Based Tools

Tools that allow states to manipulate the international trade of commodities and energy resources for strategic purposes, thereby leveraging global dependencies upon imports and exports

Sample Tools:
- Blacklists
- Boycotts and embargoes
- Dumping
- Import and export controls
- Licensing
- Most Favored Nation status
- Tariffs
- Trade sanctions

Contribution to Irregular Warfare goals:
1. **Power**: Denies the provision of key resources and technologies to adversaries
2. **Influence**: Shapes the perceptions of other states by rendering them dependent upon goods, services, and energy resources
3. **Legitimacy**: Allows states to enforce norms through multilateral sanctions regimes

Capital-Based Tools

Tools that enable states to control endowments of capital and financing, thereby depending upon a state's ability to control outbound investment and its centrality to the global financial system

Sample Tools:
- Asset freezes
- Economic assistance
- Encouraging private capital exports or imports
- Financial sanctions
- Investment guarantees
- Provision or withdrawal of assistance
- Withholding dues to int'l orgs

Contribution to Irregular Warfare goals:
1. **Power**: Empowers partner economies while undermining those of adversaries
2. **Influence**: Shapes the perceptions of other states by rendering them dependent upon capital, finance, and aid
3. **Legitimacy**: Builds and enforces norms and rules through international institutions

Domestic Policies

Tools that increase a state's economic strength and prosperity by improving domestic market features, providing more leverage and greater attractive power in achieving strategic goals

Sample Tools:
- Controlling energy and commodities
- Fiscal policy
- Providing public goods
- Industrial policy
- Monetary policy
- Regulation
- Funding R&D

Contribution to Irregular Warfare goals:
1. **Power**: Builds the size and strength of a state's economy, allowing for greater defense budgets and tech innovation
2. **Influence**: Increases a state's real and perceived leverage over key trade and capital relationships
3. **Legitimacy**: Bolsters attractive power

Tools of Economic Statecraft

Nevertheless, scholars continue to debate the potency of these tools.

One school of thought, economic liberalism—whose ideology rests upon capitalist principles—expects that the magnitude of the economic incentives or punishments propels policy success (or failure). From this perspective, prosperity is a driving force for state behavior. Thus, liberal economic theory successfully explains the allure and power of the Bretton Woods institutions.

A second theory, realism (including mercantilism), is skeptical of the role of economic factors' ability to persuade states to change behavior. Instead, security factors reign supreme, and economic tools are only important insofar as they affect vital national security interests. Thus, although less capable of explaining the power and influence of international economic institutions, realism offers a compelling explanation of the return of strategic competition and the ways in which the primacy of that competition has subordinated economic imperatives.

Finally, a conditional approach, with which this book aligns itself, asserts that economic statecraft can work under certain conditions.[11] Though economic factors are important, they are subordinate to national security imperatives. Therefore, rather than use economic means solely for prosperity, US policymakers should employ them to achieve the broader geopolitical goals of power, influence, and legitimacy, as discussed in the next section. Moreover, given their importance as a means of national security, the US must better integrate economic tools to win without fighting and ensure they complement other tools of statecraft.

Economics and Power, Influence, and Legitimacy

This section explores the relationship between economics and the objectives of IW, which are to gain power, legitimacy, and influence.

Economic Power

Just as technological and economic resources underpin military power, they are also the foundation of economic power.[12] A state's ability to change the behavior of allies and adversaries depends upon both the absolute and relative magnitude of its economic resources. Indeed, "Like all power, national economic power is relative."[13] The strength of a state's economy offers it outsized power to coerce states. For example, in April 2020, Australia called for an independent inquiry into the origins of the COVID-19 pandemic. In response, China employed various economic measures against Australia, such as tariffs on its key exports—in May 2020, China imposed an 80% tariff on Australian barley and banned the purchase of Australian beef.[14]

Power is conveyed through the size and strength of a state's economy inasmuch as it can spend relatively more on its national security capabilities than others. Larger coffers have the potential to fund greater numbers of troops and larger investments in new technologies, equipment, and training. For example, in 2022 the US authorized $740 billion in defense spending,[15] which is more than 38% of worldwide military expenditures and over three times that of the next largest military, China[16] (though recent evidence suggests PRC defense funding is far greater than widely reported).[17] This provides substantial leverage over other states. At the same time, power depends not only on the magnitude of state spending but also on how states spend their resources. Disruptive technologies and new approaches to warfare may allow countries that spend less money to develop profound asymmetric capabilities.

However, latent economic strength is not the only determinant of economic power—a state's capacity for converting potential to actual economic power is also essential.[18] A state must craft domestic policies to build economic capacity and strength and to insulate itself from shocks and crises, which can be devastating to economic power. States can subsequently use their relative economic power to wield trade-based and capital-based tools in the international arena to serve various objectives. This could include weakening or strengthening another state, threatening to shut off valuable markets, preempting sources of supply, and threatening to halt or reduce investment and economic assistance.[19]

Economic Influence

In contrast to power, economic influence can be used to sway perceptions, beliefs, and values. A state can increase its influence by pursuing economic relationships, thereby rendering other states more dependent and less autonomous. China's BRI, is a case study of using economic relationships to influence states. By providing large infrastructure investments coupled with assistance and loans, China has significantly grown its global presence and influence.[20]

Economic influence can facilitate modification of state behavior; the more a state depends upon another state for trade and capital, the likelier it is to align its behavior with the more powerful state's interests and objectives.[21] For example, tools such as sanctions are most effective when the levying state has a significant trade relationship with the sanctioned state or when the state advocating for sanctions has influence in international organizations.[22] Economic inducements may also be employed to sway a state's values. For decades Panama had recognized Taiwan's sovereignty but reversed this position in 2017, cutting off diplomatic ties and establishing more formal relations with China. This dramatic change in Panama's position has been widely attributed to Chinese economic inducements.[23]

Economics and Legitimacy

Finally, there is the question of the relationship between economics and legitimacy. Legitimacy in democracies often rests upon the processes of competitive elections.[24] However, in autocratic or communist regimes, legitimation is based largely upon the social and economic benefits a state can provide.[25] Recent research on China acknowledges that beyond merely improving well-being through economic means, citizens also care about the fairness of market interventions and form judgments about the state's effectiveness in carrying out economic policies.[26] This implies that, in some ways, inclusive political processes may insulate democratic institutions from the need to deliver economically more than autocratic regimes.

Foreign economic policies can contribute to or detract from a state's legitimacy. Chinese overseas investment and infrastructure development have significantly contributed to the PRC's influence and legitimacy, challenging American economic legitimacy in geostrategic areas (such as the Gibraltar-Suez route across the Mediterranean) and forcing the US to react instead of taking the economic initiative.[27] US support of and commitment to global economic institutions have been foundational to American legitimacy.

Yet, their failure to produce results and the requirements to uphold certain international norms—such as universal human rights and environmental protection—have provided an opening for revisionist powers to exploit. Likewise, when economic tools, such as embargoes or sanctions, fail to achieve their intended objectives, legitimacy suffers. For example, the longstanding US embargo on Cuban goods has failed to achieve its intended purpose—to undermine the regime and starve it of its resources.[28] This failure has also harmed US legitimacy. The UN General Assembly has called for the US to end the embargo 31 times passing a non-binding resolution approved by 187 countries.[29] Given that international economic institutions have been foundational to global American legitimacy, the US must take care to not risk legitimacy with the overuse of non-effective tools.

SNAPSHOT OF PROMINENT ECONOMIC TOOLS

This section assesses how the US has used and misused economic tools —trade-based, capital-based, and domestic policies—to achieve power, influence, and legitimacy.

Trade-Based Tools

The regulation of foreign trade has historically been a tool of statecraft. Most often, policymakers threaten the reduction or elimination of trade as a punishment—such as the EU embargo on imports of most Russian crude oil in 2022,[30] or UN sanctions on al-Qaeda and the Taliban.[31] However, trade can also be an inducement to cement alliances (such as the US-Morocco Free Trade Agreement) or promote détente between adversaries (the 2020 Abraham Accords between Israel and several Arab states included the potential for future free trade agreements). The tools can also be used to build power but, when overused, can negatively impact legitimacy. All can be used as part of an IW campaign to diminish competitor power, influence, and legitimacy. This section examines the role of export controls, tariffs, and trade sanctions.

Export Controls

The US began using export controls as a formal policy tool in the early twentieth century. The Export Control Act of 1940 established the first systematic control regime. During the Cold War, the US used export controls to limit adversaries' access to both military and dual-use technologies. As concerns over strategic competition have expanded, export controls have become a policy tool of choice for Washington, particularly vis-à-vis China. The US employs them both to target global companies and, as the Commerce Department continues to implement the Export Reform Act of 2018, to limit trade in a greater range of products and technologies—especially those that are considered strategic, like AI.[32] Despite this significant effort to use export controls for national security purposes, there has not been a sufficient expansion in resources and personnel to adequately administer and enforce these controls.[33]

Tariffs

Domestically, the president is permitted to impose import restrictions on goods that are deemed a threat against critical domestic industries, thus harming national security;[34] the Department of Commerce determines which imported goods pose a threat to US national security.[35] Internationally, Article XXI of the GATT allows WTO members to take measures to protect essential security interests.[36] Such tariffs in the US often affect imported steel and aluminum,[37] as well as automobiles.[38] But tariffs create market inefficiencies and, therefore, can backfire.

For example, during the Trump administration, the increased use of tariffs started a trade war.[39] In June 2018, the US imposed tariffs of 25% on $34 billion worth of Chinese imports. The Chinese retaliated with their own tariffs of 25% on 545 goods originating in the US.[40] This led to tit-for-tat retaliation in August and September 2018. Fortunately, the countries reached a truce in January 2020. US tariffs on Chinese imports, however, did not produce the desired results. US companies and consumers were forced to bear the burden of the tariffs,[41] and the trade deficit has continued to grow even as China's economy remains relatively unscathed in the short term.[42] Tariffs must be considered carefully, acknowledging the risks to one's own economy before deploying them.

Trade Sanctions
Economic sanctions refer to the deliberate withdrawal, or threat of withdrawal, of trade relations between states.[43] At their most punitive, this includes boycotts (refusing to buy goods or services from a supplier) and embargoes (fully ending trade in goods and services to a particular buyer). Sanctions aim to achieve a political goal by imposing costs on the target country, denying benefits from international trade, and limiting access to markets. The underpinning economic theory assumes that actors are rational and perceive the costs of such actions; however, it does not account for human behavior, which often undermines their effectiveness.[44]

Political goals vary widely across sanctions regimes, but generally, the more ambitious the goal, the more difficult it is to achieve.[45] For example, sanctions alone are unlikely to lead to regime change since the costs often pass over the regime onto the population. Moreover, sanctions can also fail if they unify the target country in support of its government, if they prompt powerful or wealthy allies of the target country to provide greater support,[46] or if they alienate the allies and domestic businesses of the levying state.[47] Accordingly, there is a voluminous literature that argues sanctions have a limited capacity to achieve their intended objectives.[48] Not only is their track record poor, but even when they are effective, the period in which they have any impact is brief[49]—the first and second years account for 55% of successful sanctions cases, followed by a steep decline.[50]

Despite these critiques, sanctions occasionally work, especially when appropriately scoped. When policy goals are modest, sanctions have a much higher chance of success. The likelihood of success also increases when sanctions target allies and trading partners, which are often more easily coercible than adversaries, because democracies are more susceptible to economic pressure than autocratic regimes. Additionally, if a state applies maximum pressure immediately, instead of incrementally[51] and acts in coordination with coalitions, sanctions are more likely to succeed. Since the Cold War, but especially since 9/11, the US has employed sanctions too often and ineffectively, choosing unilateral action against non-democratic regimes and overly ambitious concessions such as regime change or nuclear disarmament. When used appropriately, sanctions can be integral to an IW campaign and erode competitor power, as the recent multilateral sanctions on Russia in response to its invasion of Ukraine have shown.

Capital-Based and Financial Tools
Capital includes cash, liquid investments, raw materials, and finished goods while finance encompasses the provision of a banking system, the circulation of money, granting credit, and making investments.[52] As cross-border economic flows have increasingly become financial, rather than physical, capital-based tools are increasingly important. For example, in the 1970s, "90 percent of all cross-border flows were trade-based; in 2014, 90 percent were financial."[53]

Today, the US exerts considerable economic power through global capital markets. To achieve this, the Treasury Department exercises wide authority to target the financial assets of individuals, groups, or countries deemed national security threats.[54] Like trade-based tools, the carrot is often as useful as the stick, particularly in supporting allies and partners and countering the threats posed by competitors. Private sector investment overseas and access to governments through economic assistance also offer opportunities for influence.[55] This section will examine financial sanctions, asset freezes, investment screening, and economic assistance.

Financial Sanctions
The centrality of the dollar to the international monetary system facilitates America's outsized influence over financial transactions. Since 9/11, the US has made it more difficult for financial institutions to engage in dollar transactions with sanctioned governments, companies, or individuals.

> US and foreign banks need access to US dollars to function; even the implicit threat of being denied such access has made most banks in the world reluctant to work with sanctioned entities, effectively expelling them from the global financial system.[56]

Many sanctions regimes aim at creating financial hardship, including withdrawal of development aid, expulsion from SWIFT, and freezing assets such as central bank foreign currency holdings.

The most remarkable part of US unilateral financial sanctions is their application to non-US citizens or companies—referred to as "secondary sanctions." The 2012 National Defense Authorization Act (NDAA) authorized the Treasury Department's Office of Foreign Asset Management (OFAC) to cut off non-US citizens' or companies' access to the American market. When a non-US citizen or company is found trading with entities on the sanctions blacklist, OFAC can take a range of measures, from limiting their commercial activities to completely prohibiting US market access. To delist from the secondary sanctions list, the offending entity must pay a large financial settlement.[57] The size of the US market allows OFAC to make a credible threat of significant financial damage to a company or individual; however, the exploitation of secondary sanctions has incentivized some states to reduce their dependence on the dollar.[58]

Asset Freezes
Asset freezes are government-imposed administrative measures to target individuals or entities.[59] Given the challenges of sanctions, many policymakers favor asset freezes, which seem more precise. For example, the US government has blacklisted many senior Venezuelan government officials, including the president and the last two vice presidents. However, this policy has had unintended consequences. It has impacted Venezuela's ability to buy agricultural imports, which negatively affects the broader population more than the targeted government officials. Similarly, when the US blacklisted Venezuela's national oil company, which generates 95% of the country's export revenue, the overall economy was affected.[60] This denied the country access to billions of US dollars worth of gold reserves, trade credits worth more than $3 billion, and the oil company's net assets, estimated at over $5 billion.[61]

Asset freezes can impose immense costs for those placed on a sanctions list. Conversely, asset freezes can uniquely bolster US legitimacy—the Global Magnitsky Human Rights Accountability Act of 2016 authorizes the Treasury Department to sanction foreign government officials accused of human rights abuses, even when there are no sanctions against the country.[62] Unfortunately, the number of individuals and companies on these lists has been extremely difficult for banks to manage.

Investment Screening
Governments increasingly use enhanced regulations to review and block cross-border mergers and acquisitions that threaten to provide other states with access to strategic resources and technologies. A notable example of a cross-border merger that raised US national security concerns involves the attempted acquisition of Qualcomm, an American semiconductor

and telecommunications equipment company, by Broadcom, a Singapore-based (later redomiciled to the US) technology company.[63] The US government, through the Committee on Foreign Investment in the United States (CFIUS), [64] expressed concerns over the acquisition. The primary worry was that the takeover could weaken Qualcomm, a leader in developing 5G technology, and consequently cede US leadership in this critical area to foreign competitors, particularly China.

In 2018, Congress expanded CFIUS's jurisdiction through the Foreign Investment Risk Review Modernization Act, giving it greater powers to review minority investments and expanding the scope of investments requiring review.[65] This move primarily aims at preventing Chinese firms from exploiting US capital markets to acquire technology.[66] Now, even minority stakes in US companies are subject to scrutiny,[67] particularly those specializing in critical technologies like AI, robotics, augmented and virtual reality, biotechnology, and new financial technology. As a result, CFIUS has become more assertive in reviewing prospective investments and has vastly expanded its reviews of completed investments not previously covered, requiring divestment or other mitigation measures. These changes have largely been considered successful in balancing the promotion of investment while protecting US national security.[68]

Economic Assistance
One of the crucial objectives of US foreign economic assistance is to support national security.[69] Its use to increase strategic influence is one of the most straightforward examples of economic statecraft. "Foreign economic assistance is often used to promote a donor country's interests by helping to stabilize or develop a recipient state's economy or otherwise to influence its social, political, or military actions."[70]There is also the threat of withdrawing economic assistance, predicated on recipient behavior. China's BRI, which offers loans and grants for infrastructure projects, illustrates how incentives provide extraordinary leverage on recipient countries, either by offering finance or withdrawing it.

Economic assistance contributes to influence and legitimacy when local populations and political elites realize financial benefits. As chapter 4 notes, the most well-known instance of US economic assistance was the Marshall Plan to rebuild Europe after World War II: the billions of dollars not only alleviated human suffering but also provided a bulwark against Soviet expansion. Besides the Marshall Plan, and the multi-decade US President's Emergency Plan for AIDS Relief there are few clear instances of successful economic assistance programs. As a negative example, American aid to anti-communist regimes to Latin American governments during the Cold War was associated with brutal violence and damaged US legitimacy.[71]

Post-9/11, US foreign assistance includes loans, grants, and technical assistance to gain cooperation on national security goals and build capacity in failed or failing states. In addition, international organizations such as the IMF, the IFC, and the World Bank provide subsidized loans or loan guarantees for development projects to stabilize fragile states and enable trade.[72] Yet, few of the recipient states have successfully transitioned from disorder to stability. Nevertheless, capital-based and financial tools provide many options for supporting an IW campaign, from supporting allies and partners with economic assistance to freezing assets of nefarious actors and restricting adversary access to strategic resources.

Domestic Policies
In the US, markets drive most economic decisions, although the government also plays an important role. Besides providing public goods, such as national defense, the US government regulates the environment, defines and protects property rights, and attempts to make markets more competitive. Domestic policies have the potential to set the foundation by which states create economic power and increase their competitiveness. An array of domestic tools that increase economic competitiveness and resilience include the provision of public goods, protection of property

rights, research and development, monetary and fiscal policy, policies governing energy and commodities, and industrial policy.

Provision of Public Goods

The provision of public goods can increase a state's economic power and legitimacy in various ways. Public goods include infrastructure (transportation, financial, cyber, water, energy), education, public health, and law enforcement—all services that benefit society and make an economy more competitive. A well-educated workforce is essential to economic growth, and the resultant human capital can be harnessed during crises. Investments in education and training improve labor productivity and foster innovation. Likewise, a healthy and safe workforce is more productive. Public goods like advanced research facilities, transportation networks, and digital infrastructure can create relative power by giving it a competitive edge in the global market. Meanwhile, public safety, environmental protection, and social welfare programs contribute to social cohesion and stability, key components of legitimacy.

Property Rights

Property rights play a crucial role in building economic power and legitimacy. Secure intellectual property and land rights provide individuals and businesses with the confidence to invest knowing that their proceeds will be legally protected. This fosters innovation as companies are more likely to develop new products and technologies when they are assured of benefiting from their work and the resources they committed adding to state power and influence. A government that effectively protects property rights gains legitimacy in the eyes of its citizens and the international community.

Research and Development (R&D)

During the Cold War, federal R&D funding contributed significantly to the prosperity, security, and primacy of the US. The US government funded basic and applied research projects, providing two-thirds of research dollars in 1956.[73] This led to the development of vital security capabilities, weapons and space technologies, as well as provided significant private sector benefits. For example, nearly every component of Apple's iPod and iPhone was created due to government funded R&D.[74] Today, the private sector is the primary source of innovation and R&D funding (70 percent of funds), but new technologies increasingly blur the lines between military and civilian use.[75]

Monetary and Fiscal Policy
A state can translate monetary policy tools into influence through the global footprint of its currency, the ability to borrow funds at a low cost, and the ability to impact another state's cost of borrowing.[76] The valuation, or devaluation, of a state's currency has the potential to provide a relative advantage—indeed, if a trading partner deliberately pursues policies to weaken its own currencies, it could gain an unfair advantage in international trade. For example, China kept its currency artificially cheap for decades to give its exporters an unfair edge.[77]

Monetary and fiscal policy are particularly potent for the US, given the centrality of the US dollar in the world economy. "The dollar is America's superpower,"[78] accounting for about 59% of foreign exchange reserves,[79] half of international trade, 40% of international payments, and 85% of foreign exchange transactions.[80]

> The country that gets to print the world's primary currency...is in a very powerful position, and debt that is denominated in the world's reserve currency...is the most fundamental building block for the world's capital markets and the world's economics.[81]

However, the power of the dollar has incentivized states to circumvent its enormous influence by developing alternatives. For example, China is building a SWIFT alternative to rival the dollar;[82] India is settling its oil purchases from Russia in nondollar currencies; and Saudi Arabia has considered pricing its oil in RMB.[83] To maintain economic primacy, the US must recommit to sound monetary and fiscal policy to extend the power of the dollar. The loss of currency status has historically coincided with the loss of great-power status.[84] Concurrently, the US government must reconsider how frequently it uses economic tools (especially sanctions to punish individuals, groups, and states) that encourage the development of viable dollar competitors.

National Policies Governing Energy and Commodities
Ensuring a constant flow of energy is essential for the functioning of economies.[85] The US and Russia are major producers of both oil and natural gas, providing both countries the ability to exert influence over world hydrocarbon markets. In 2021 almost 40% of Europe's gas was provided by Russia,[86] giving it the potential to exercise leverage and potentially divide European governments as it invaded Ukraine. China, though a major oil producer, imports the majority of its oil and is dependent on external sources for its natural gas consumption.[87]

The US has recently ramped up its energy production—especially that of liquified natural gas (LNG)—to increase energy diversification, thereby limiting its own exposure and that of its allies to international energy price changes.[88] However, inadequate infrastructure has prevented greater expansion of LNG use, even in Europe. If LNG is to be a long-term competitor to Russian gas, the US will need to invest more to support infrastructure expansion.

Since the Soviet Union's dissolution, Russia has threatened to suspend or has actually suspended gas supplies to parts of Europe more than fifty times.[89] However, the tendency to use energy and commodities for political use is not uniquely Russian.[90] China has also used energy and commodities as geopolitical instruments. In 2010, it banned exports of rare earth elements to signal its dissatisfaction with neighboring countries' policies.[91]

Industrial Policy

Though market failure is usually the rationale for state interference in the economy, governments may also choose to intervene to protect specific industries and develop critical technologies. Since World War II, the US has maintained a complex, constantly evolving relationship between the aerospace and defense industry on the one hand and the acquisition personnel of the DOD on the other. Despite an underlying capitalist aversion to industrial policy, the monopsonist nature of the American defense sector has imbued the DOD with the power to influence this market through its demand signals and incentive structures.[92] Since the 1990s, the US government has employed acquisition reform, regulation of defense firms and consolidation of the defense industry.[93] However, it still struggles to strike the right balance of innovation and competition to provide desired capabilities at reasonable cost with secure sources of supply.

Recent technological developments further complicate industrial policy. Whereas past military innovations like stealth and precision-guided munitions belonged neatly in the defense sector and sprouted from public-private collaboration, today's most transformative technologies like AI, autonomous technology, and 5G are often dual-use and mature primarily in the private sector. As a result, the expansion of industrial policy to protect and leverage these industries is often difficult.

Fears regarding China's global export of 5G networks exemplify this. Although the US has used government intervention to bolster nascent firms to compete in the 5G market, it has still struggled to rival cheaper Chinese companies. Further, most markets outside the US have fewer security concerns about Chinese access to individual user information.[94] Moreover, in the wake of Google's withdrawal from Project Maven (a DOD program to improve analysis of drone footage which sparked public criticism) and Microsoft's concern about augmented reality headsets (the purpose of the headsets was initially to train soldiers, but employees were uneasy about turning the technology into a weapon), some technology firms are wary of working with the government to support military missions.[95] Fortunately, many technology firms do want to work with the US government; nonetheless, the USG and its subordinate agencies must improve their approach to industrial policy to facilitate this collaboration.

RECOMMENDATIONS

Economic power, influence, and legitimacy underpin military strength and contribute directly to a state's power in the international system. Trade and capital-based tools, in combination with domestic policies, should be used more extensively to diminish the power, influence, and legitimacy of competitors—the objective of IW. What follows are a few ways the US could more effectively convert its economic capability into relative power, influence, and legitimacy.

Expand the Treasury Department's Size and Require an Economic Statecraft Strategy

The Treasury Department has increasingly served national security policymaking, yet has not grown commensurate with the dramatic change in demand. Furthermore, it lacks an office dedicated to strategic planning akin to the DOS policy planning staff. To truly integrate economic statecraft as a core part of US national security, the Treasury Department, in conjunction with USAID, the Commerce Department, and the US Trade Representative, should be required to develop and implement a holistic Economic Competition Strategy every four years. This requirement should be linked to the development and publication of the NSS.

This effort requires a new office and additional personnel at the Treasury Department and the other agencies involved. Barry Pavel and Daniel Egel go so far as to suggest the establishment of a civilian Joint Chiefs of Staff, which would include the Secretaries of Commerce, State, and Treasury Departments, to elevate economic statecraft.[96] Without dedicating the resources and elevating economic statecraft as a priority, there is little hope of harnessing America's full economic potential.

Reframe Domestic Issues as Essential Components of National Security

When George Kennan wrote the Long Telegram in 1947, he assumed the US economic model would continue to succeed of its own accord. Without a concerted effort to address the twenty-first-century weaknesses of its economy, the US leaves itself vulnerable.[97] The best way to shore up US economic power, influence, and legitimacy is to reinvest in the domestic underpinnings of a strong economic and democratic society. This includes cutting excessive debt, electing a legislative branch capable of routinely and predictably passing a budget, improving underperforming public goods, financing underfunded research and development, and eliminating economically irrational immigration policies. Though admittedly, current political realities hamper consensus, reframing these issues as vital to American national security may facilitate much-needed change.

Avoid Protectionism

The US government's quest for global free trade dates back to the 1920s. Yet, beginning with the Trade Expansion Act of 1962, several industries sought safeguards against foreign competition, slowing the pace of multilateral liberalization.[98] More recently, the Trump administration implemented the biggest reversal of multilateral trade liberalization since the early twentieth century, which the Biden administration has quietly maintained.[99] Yet, Americans have overwhelmingly gained from trade liberalization; trade supports nearly 39 million American jobs.[100]

In the 25 years since the establishment of the WTO, the US annual GDP has increased by $85 billion, a larger positive impact than that of any other WTO member.[101] Though Congress and the president generally work together to negotiate and implement US trade agreements, Congress has periodically delegated limited authority to fast-track trade agreements to the president; this is called Trade Promotion Authority (TPA). The last TPA expired in 2021, and the Biden administration has not yet requested its renewal.[102] The US should recommit to using trade agreements as a carrot instead of over-relying on negative tools of economic statecraft.

Reinvest in Multilateralism

Multilateralism, while difficult, boosts US influence and legitimacy. Moreover, the tools of economic statecraft work best when implemented multilaterally. From trade to sanctions to economic development, multilateral organizations and agreements are simply more impactful than acting alone. Though COVID-19 exposed vulnerabilities from overreliance on global supply chains, and China's coercive use of economic tools similarly negatively impacts American allies and partners, this should not be used as an excuse to undermine or withdraw from international economic institutions. At least part of the story of the world's tremendous economic growth and the trend of poverty reduction over the last fifty years has been the commitment and support of the Bretton Woods institutions and global norms around economic integration and connectedness. The US should not only recommit to multilateralism where it can, but publicly advocate for the ideals that underpin them —both of which contribute to US legitimacy and the international organizations the US helped create.

Reconsider the Overuse of Sanctions

Though the US is often eager to impose sanctions, it is an arduous and highly bureaucratic process to withdraw them. Popular to use, sanctions require political commitment to remove and consequently difficult to end.[103] Further, if sanctioned states do not believe that sanctions can be lifted, there is little incentive to comply with the demands of the sanctioning state. Finally, sanctions harm the economies of both the sanctioning state and the sanctioned state.

Economic coercion is best deployed when the coercing state is extremely clear about the conditions under which sanctions will be threatened, enacted, and lifted.[104] Sanctions could be enacted with an annual expiration date, forcing regular reconsideration of their effectiveness. Building in expiration dates and reporting requirements (perhaps quarterly or every six months) has worked well with UN Security Council Resolutions. A similar construct could be developed for US sanctions.

Reinvest in and Incentivize R&D

During the Cold War, Russia's demonstration of its superior satellite technology spurred an era of technological innovation. The government founded the Defense Advanced Research Projects Agency (DARPA) and National Aeronautics and Space Administration (NASA), expanded the National Institutes of Health (NIH), centralized DOD science and technology policy, and created the Small Business Investment Companies (SBIC) program. To date, China's high-tech capabilities have not triggered a similar reaction.[105]

Reinvesting in technology, particularly AI and 5G, is essential to competing with China. The 2022 CHIPS and Science Act is an essential first step to encouraging emerging technology research, reducing critical technology dependency on China, and incentivizing the expansion of US domestic semiconductor production.[106] At the same time, the semiconductor industry is but one sector that requires substantially more R&D funding. To truly compete, the US must mobilize a whole-of-society approach that draws on government support for the private sector and academia.

CONCLUSION

One of the most powerful tools the US possesses is its economic strength, giving it enormous potential to use it as a core component of IW as well as to withstand the era of crises. The centrality of the dollar, the power of US-created international financial institutions, and the size of the American market provide enormous advantages and enable the use of economic statecraft to gain relative power, legitimacy, and influence. The weaknesses in the current American approach can be fixed through a better understanding of these advantages and the more deliberate wielding of the tools of economic statecraft..

Notes

1. Baldwin, *Economic Statecraft*.
2. Blackwill and Harris, *War by Other Means*; and Meese et al., *American National Security*, 288.
3. Kennedy, *The Rise and Fall*; and Wigell, "Conceptualizing Regional Powers."
4. Farrell and Newman, "Weaponized Interdependence"; and McCormick, Luftig, and Cunningham, "Economic Might."
5. Zarate, *Treasury's War*.
6. Matthijs and Meunier, "Europe's Geoeconomic Revolution," 168; and Walentek, "Economic Peace Revisited."
7. Lagarde, "Updating Bretton Woods."
8. Bachand, "International economic institutions."
9. Knorr, *The Power of Nations*.
10. Hans Morgenthau identified nine elements of (potential) national power; geography, natural resources, industrial capacity, military preparedness, population, national character, national morale, quality of diplomacy, and quality of government. See Morgenthau, *Politics Among Nations*.
11. Blanchard and Ripsman, *Economic Statecraft and Foreign Policy*, 373.
12. Knorr, *The Power of Nations*.
13. Knorr, 102.
14. "China punishes Australia for promoting an inquiry into covid-19."
15. Shane and Gould, "Congress passes defense policy bill with budget boost, military justice reforms."
16. Tian et al., "Trends in World Military Expenditure, 2022," 2.
17. Recent estimates of China's military budget may be as high as $700B, equivalent to roughly 4 percent of the PRC's GDP. Eaglen, "China's Real Military Budget Is Far Bigger Than It Looks."
18. Knorr, *The Power of Nations*, 46.
19. Knorr.
20. McBridge, Berman, and Chatzsky, "China's Massive Belt and Road."
21. Gartzke and Westerwinter, "The complex structure of commercial peace."
22. International Crisis Group, "Sanctions, Peacemaking and Reform."
23. Calamur, "Panama's Decision to Cut Ties with Taiwan."
24. White, "Economic Performance and Communist Legitimacy."

Tools of Economic Statecraft 137

25. White.
26. Eaton and Hasmath, "Economic Legitimation in a New Era."
27. Woo and DMichaels, "China Buys Friends."
28. Griswold, "Four Decades of Failure."
29. "U.N. votes to end US embargo."
30. Berman and Siripurapu, "One Year of War in Ukraine."
31. United Nations Security Council, "United Nations Security Council Consolidated List."
32. Sullivan, "Remarks by National Security Advisor."
33. Allen, Benson, and Reinsch, "Improved Export Controls."
34. Chatzky and Siripurapu, "The Truth About Tariffs."
35. Fefer, "Section 232 of the Trade Expansion Act of 1962."
36. World Trade Organization, The General Agreement on Tariffs and Trade, Article XXI.
37. US Department of Commerce, *The Effect of Imports of Steel.*
38. Fefer et al., "Section 232 Investigations."
39. Bown, "US-China Trade War Tariffs."
40. Hua and Zeng, "The US-China Trade War."
41. Hua and Zeng.
42. Khalid, "Biden kept Trump's tariffs on Chinese imports. This is who pays the price." Though the analysis of the impacts of the tariffs on China were relatively small in the short run, China's long-term economic outlook might be vastly different. Bown, "Four Years."
43. Hufbauer et al., *Economics Sanctions Reconsidered*, 3.
44. Smeets, "Can Sanctions Be Effective?"
45. Smeets.
46. Hufbauer et al., *Economic Sanctions Reconsidered.*
47. Hufbauer et al..
48. Hufbauer et al.; Drezner, "Sanctions Sometimes Smart"; Smeets, "Can Economic Sanctions Be Effective?"; Blanchard and Ripsman, *Economic Statecraft*; and Drezner, "The United States of Sanctions."
49. Smeets, "Can Economic Sanctions Be Effective?"
50. Hufbauer et al., *Economic Sanctions Reconsidered.*
51. Hufbauer et al., 166.
52. Katz, "Waging Financial Warfare."
53. Blackwill and Harris, *War by Other Means*; and Steil and Litan, *Financial Statecraft.*
54. Temple-Raston and Rishikof, *The National Security Enterprise*, 162–181.
55. Knorr, *The Power of Nations.*

56. Drezner, "The United States of Sanctions."
57. Gorny, "Primary and Secondary Sanctions Explained."
58. McDowell, "Financial Sanctions and Political Risk."
59. Oldfield, "The Challenges of Asset Freezing."
60. Gordon, "The Not So Targeted Instrument."
61. Weisbrot and Sachs, "Economic Sanctions."
62. Weisbrot and Sachs.
63. "Timeline: Broadcom-Qualcomm saga."
64. Tarbert, "Modernizing CFIUS."
65. Khanapurkar, "CFIUS 2.0."
66. Khanapurkar.
67. Aggarwal and Reddie, "Economic Statecraft in the 21st Century."
68. Arasasingham and DiPippo, "Evaluating CFIUS in 2021."
69. Morgenstern and Brown, "Foreign Assistance."
70. Meese et al., *American National Security*, 301.
71. Lawson, "Does Foreign Aid Work?"
72. Meese et al., *American National Security*, 304.
73. Sargent, "US Research and Development."
74. Mazzucato, *The Entrepreneurial State*.
75. McCormick, Luftig, and Cunningham, "Economic Might."
76. Blackwill and Harris, *War by Other Means*, 75–76.
77. Phillips, "China's Currency is Weakening."
78. Zakaria, "The Dollar is Our Superpower."
79. International Monetary Fund, "Currency Composition."
80. Nelson and Weiss, "The US Dollar."
81. Dalio, *Principles for Dealing with the Changing World Order*.
82. Cohen, *Currency Statecraft*.
83. Zakaria, "The Dollar is Our Superpower."
84. Dalio, *Principles for Dealing*.
85. Kurecic, "Geoeconomic and Geopolitical Conflicts."
86. BP, "Statistical Review of World Energy 2022."
87. Energy Information Agency, "China imported record volumes of oil."
88. Diaz et al., "US Energy in the 21st Century."
89. Johnson, "Putin's Gas Gambit Backfires."
90. Blackwill and Harris, *War by Other Means*, 85.
91. Blackwill and Harris.
92. Siriprapu and Berman, "Is Industrial Policy Making a Comeback?"
93. Grasso, "Defense Acquisition Reform."
94. Aggarwal and Reddie, "Economic Statecraft in the 21st Century."

95. Owen, *Lessons from the US*; and Shane, Metz, and Wakabayashi, "How a Pentagon Contract."
96. Pavel and Egel, "A Civilian US 'Joint Chiefs'?"
97. Atlantic Council, "The Longer Telegram."
98. Chase, "The United States and Multilateral Trade," 106.
99. Bacchus, "Biden and Trade at Year One."
100. Trade Partnership Worldwide, *Trade and American Jobs*, 2.
101. Felbermayr, Yotov, and Yalcin, "The World Trade Organization," 33.
102. Akhtar et al., *US Trade Policy*.
103. Drezner, "The United States of Sanctions."
104. Drezner, "Sanctions Sometimes Smart," 96–108.
105. Weiss, "Re-Emergence of Great Power Conflict."
106. Yang, "Can the CHIPS Act?"

CHAPTER 7

TOOLS OF INFORMATION STATECRAFT

> Information is now the world's most consequential and contested geopolitical resource.[1]
> Eric Rosenbach and Katherine Mansted

INTRODUCTION

Information statecraft is a key component of IW and strategic competition.[2] Though the US and the Soviet Union both successfully weaponized information during the Cold War, the internet and social media have fundamentally changed how information is produced, communicated, and distributed.[3] This has, in turn, provided greater opportunity for authoritarian governments like those of China and Russia to use information maliciously, elevating its significance as an essential tool for strategic competition.

China and Russia deliberately amplify false or misleading information, spread conspiracy theories designed to deflect blame for their own wrongdoing, undermine the prestige of the US, and cast doubt on the notion of objective "truth."[4] This trend is alarming for the US, as access to truthful information and its unencumbered exchange among citizens are foundational to democratic principles. A strong democracy requires high-quality news from independent media, a pluralistic climate of opinion, and the ability to negotiate public consensus.[5] Democracy also depends on the idea that truth is knowable, and citizens can discern it.[6] These principles are currently under direct attack.

Information and its manipulation pose a significant threat to the US, which will only grow as technology advances. As Frederic Filloux explains, "what we see unfolding right before our eyes is nothing less than Moore's Law applied to the distribution of mis-information: an exponential growth of available technology coupled with a rapid collapse of costs."[7] To help understand these interrelated phenomena, this chapter provides an overview of the terms used to describe information, its types, and its uses, as well as how information is used to impact power, influence, and legitimacy. Next, it evaluates how the US has used and is using information to gain relative power, influence, and legitimacy. The chapter concludes with recommendations for adapting information statecraft and incorporating it into IW campaigns in an era of crises and strategic competition.

Contemporary Information Statecraft

Global communication technologies can rapidly disseminate misinformation and disinformation, which are particularly potent in spreading chaos and doubt among targeted communities.[8] Though fabricating information is not new, the complexity and scale of information pollution in a digitally interconnected world present unprecedented challenges.[9] The rise in internet saturation and social media dependence has broadened and deepened individual, community, and state vulnerability to disinformation.[10] Moreover, the absence of human editors in social media news feeds makes it easy for political actors to manipulate and deceive the voting public.[11] Social media algorithms, in particular, can be co-opted for malicious activities: interfering with elections, manipulating public opinion, or even toppling regimes.[12]

The 2016 Russian attack on the US presidential election to undermine the legitimacy of American democracy proved a watershed moment.[13] Using techniques it had deployed domestically and in Eastern Europe, Russia's Internet Research Agency (IRA) began targeting US voters for misinformation as early as 2012.[14] In particular, "The IRA adapted techniques from digital advertising to spread disinformation and propaganda by creating and managing advertising campaigns on multiple platforms, often making use of false personas or imitating activist groups."[15] The IRA targeted users by race, ethnicity, and identity on social media platforms.[16]

Four years later, the COVID-19 pandemic again underscored the fragile —and easily exploitable—relationship between information and public discourse. On February 2, 2020, the World Health Organization (WHO) released a COVID-19 situation report that described the pandemic as featuring a parallel infodemic: "an overabundance of information—some accurate and some not—that makes it hard for people to find trustworthy sources and reliable guidance when they need it."[17] Multiple narratives claiming that the spread of COVID-19 was intentional contributed to this infodemic, causing an international wave of fear and suspicion as the disease spread. The existence of competing and conflicting narratives from different countries, as well as from official and non-official sources, added to the chaos circulating across social media, traditional media, and general public discourse.[18]

DEFINITIONS AND THEORIES OF INFORMATION

Definitional Debates

Information is a critical input to individual, organizational, and governmental decision-making; it can be defined as data that has been processed, organized, structured, or presented in a meaningful way to convey knowledge, meaning, or significance.[19] But how does one conceptualize the environment in which it lives? The information environment is "the aggregate of social, cultural, linguistic, psychological, technical, and physical factors that affect how humans and automated systems derive meaning from, act upon, and are impacted by information."[20] This environment comprises three components: connectivity (the physical and virtual technologies enabling communication); content (the actual information itself); and cognition (the processes by which human brains translate content into meaning).[21]

Defining and explaining manipulation of the information environment has prompted scholarly disagreement, and a variety of competing terms are used in various contexts. However, they should not be interchangeable; each requires a slightly different policy solution.[22] Therefore, this chapter suggests the following lexicon:

- **Propaganda**: The propagation of an idea or narrative with the aim of influence to further one's cause or damage an opposing one. It can be truthful information, albeit presented selectively to highlight some facts while omitting others; it may also include information discovered through illegal means.[23]
- **Public Diplomacy**: A government's efforts (including communication, information, and propaganda) to improve its reputation with people in other nations.[24]
- **Misinformation**: The spreading of unintentionally false information, which can sow divisiveness and chaos in a targeted society.[25]
- **Disinformation**: Intentionally false information or distorted claims, often spread as news stories or simulated documentary formats;[26] to mislead people.[27] Fake news, a type of disinformation, refers to news articles that are intentionally and verifiably false, designed to manipulate perceptions of reality.
- **Foreign Malign Influence**: Any hostile effort undertaken by, at the direction of, on behalf of, or with the substantial support of, a state's government with the objective of influencing the political, military, or economic policies or activities of the US government

or sub-national governments, including elections or public opinion within the US.[28]

The increasing salience of these terms reveals both the crucial and expanding role of information in the digital age and the ways in which the reliability and accuracy of information have been undermined. In national political and civil discourse, disagreement about what information is factual has created "truth decay."[29] This phenomenon is represented by four related trends: increasing disagreement about facts and their interpretation; a blurring of lines between fact and opinion; "increasing relative volume and influence of opinion and personal experience over fact; and declining trust in formerly respected sources of factual information."[30]

These trends are troubling, especially for democracies, as information and truth underpin the interaction between citizens and their government. Today, information has become both a source of wealth and a weapon to dupe the global public or to cement authoritarian rule, playing an integral role in both "strategic competition and military conflict in the first two decades of the twenty-first century."[31] This sets the context for understanding the differing terms for the use of information.

The hierarchy of information as a tool of statecraft generally flows from the strategic level to the tactical level with the respective terms: information statecraft, influence operations, information warfare, and information operations.

- **Information Statecraft**: the broad strategic term for how a state uses information and narrative tools to achieve its strategic objectives.[32]
- **Influence Operations**: refer to whole-of-society capabilities used to erode an adversary's influence and legitimacy or to enhance one's own.[33]
- **Information Warfare**: includes the range of military and government operations to protect and exploit the information environment.[34]

- **Information Operations**: can be offensive or defense in nature and takes place at the operations level to implement an information warfare strategy.[35]

The objectives of these uses of information are generally to gain relative power, influence, or legitimacy in the eyes of foreign (but also domestic) audiences.

Information Statecraft

Information, like other elements of power, has long been used to achieve political aims. Its elevation in importance coincides with new terms for describing its use. For example, the 2017 NSS devotes an entire section to information statecraft without ever defining it. Instead, it highlights the risks of Russian and Chinese weaponization of information to "attack the values and institutions that underpin free societies, while shielding themselves from outside information."[36] This book adopts Audrye Wong's definition: "a government's use of media and information tools to spread certain narratives and alter public discourse in support of its strategic goals."[37] Information statecraft encompasses two important constituencies that are targeted or impacted in its use and misuse: a state's domestic audience (which is often targeted in authoritarian regimes but cannot be targeted in the US due to laws and democratic norms), and a state's foreign audience. Importantly, the term connotes a whole-of-government approach to gaining power, influence, or legitimacy.

Influence Operations

Influence operations aim to manipulate or influence an audience.[38] Influence operations comprise "the coordinated, integrated, and synchronized application of national diplomatic, informational, military, economic, and other capabilities in peacetime, crisis, conflict, and post-conflict to foster attitudes, behaviors, or decisions by foreign target audiences that further US interests and objectives."[39] They are part of a hierarchy of activities conducted by foreign adversaries: influence activities (single use of illegitimate techniques); influence operations (multiple coordinated activities); and influence campaigns (multiple coordinated operations across the hybrid spectrum).[40] US influence operations differ from information operations in their use of whole-of-society capabilities.

Information Warfare
Though framing the use of information as a whole-of-government endeavor is new, the use of information as an essential component of warfare has a long and storied history. Military scholars trace this trend to Sun Tzu's emphasis on accurate intelligence for decision superiority over a stronger foe.[41] Some cite the American Revolution as the first example of information warfare in the US. American rebels distributed leaflets inviting Hessian mercenaries and British common soldiers to desert, highlighting the disadvantages of the British Army and the advantages of settling in America.[42] Anti-British propaganda in American newspapers dampened domestic loyalist opposition.[43] Two and a half centuries later, the ability to control, disrupt, or manipulate the enemy's information infrastructure has become as decisive as weapon superiority in determining the outcome of conflicts.[44]

The term "information warfare" has many definitions in the literature. At its simplest, it is "a conflict or struggle between two or more groups in the information environment."[45] Robert J. Wood defines it more narrowly as "hostile activity directed against any part of the knowledge and belief systems of an adversary,"[46] whereas Jarred Prier defines it more broadly as "the use of information and communication technologies and platforms to manipulate and influence public opinion, disrupt societal harmony, spread information or disinformation, and undermine the stability and security of a nation."[47] The contention in definitions generally relates to information warfare's scope and whether it is limited to "conducting or defending against electronic attacks on computers and related information systems or whether it also includes the whole spectrum of possibilities for using information effectively in warfare and denying enemies the same ability."[48] A more recent conception, which this book adopts, views the purpose of information warfare as the pursuit of advantage over competitors by targeting an adversary state's government, military, private sector, and population as a whole-of-society endeavor.[49] This view recognizes that actors beyond the government may become agents. Indeed, citizens may function as proxies on behalf of a government with or without their knowledge.[50] Though the reference to warfare lends a militaristic connotation to the term, information warfare can be (and often is) waged during peacetime.

Information Operations
The term "information operations" is predominately used in a military context.[51] Whereas information warfare takes place at the strategic level, information operations involve using informational capabilities to support the implementation of a broader strategy at the operational level. The DOD defines information operations as "the integrated employment, during military operations, of information-related capabilities in concert with other lines of operation to influence, disrupt, corrupt, or usurp the decision-making of adversaries and potential adversaries while protecting one's own."[52] Such operations may be overt (allowing a government's production and dissemination of materials to publicly convey its values) or covert (to allow government deniability for the creation or dissemination of material). Though American information operations are generally the purview of official intelligence or military organizations, or government-contracted organizations; competitors also employ private or quasi-government actors to conduct such operations.[53]

Theories of Information Statecraft
The role of information statecraft in strategic competition is linked to the difference between how autocracies and democracies view information. Autocrats view it as a threat to domestic stability and a weapon to wield abroad.[54] Democracies, however, view open and trustworthy information as foundational to a healthy society.[55] Those views impact how the US and its competitors use information statecraft.

The US has long believed that the internet and related technologies are inherently democratizing. In 1996, Joseph S. Nye Jr. and William A. Owens suggested that America had a comparative advantage, from its status as an open society, in its ability to collect, process, act upon, and disseminate information, an edge they predicted would grow.[56] At the time, policymakers believed they could use US information resources to engage China and Russia to prevent them from becoming hostile.[57] A 1999 RAND study that suggested American information strategy should revolve around "guarded openness" remains foundational to how policymakers think about the information environment and its use in international relations.[58]

The core principles underpinning guarded openness include the belief that information is a positive force. As such, freedom of information is a right (and responsibility), making information sharing the ideal.[59] These assumptions have been enduring—for example, the US 2011 *International Strategy for Cyberspace* empowers people "to help make their governments more open and responsive" through the opportunities inherent in cyberspace,[60] while challenging authoritarian foreign regimes.[61]

Conversely, many autocrats believe that Mikhail Gorbachev's *glasnost* and *perestroika* policies contributed to the collapse of the Soviet Union by granting the media the freedom to criticize the political system and permitting civil society to participate in anti-government activism.[62] As a result, most twenty-first-century dictators pursue comprehensive control over public communication, largely banning private media and censoring the press.[63] Today, China and Russia work diligently to limit the rights of media and non-governmental organizations, seeking to dominate civil society with government-controlled organizations, including pro-regime think tanks.[64]

Russia and China are fundamentally concerned with the control of information and use various tools to undermine and discredit Western liberal norms while promoting their narratives. By targeting democratic political systems and manipulating information, they advance their foreign policy objectives without risking conventional military conflict.[65] China and Russia also push for cyber sovereignty, whereby the state would control information and data flowing through its territory.[66] This notion is completely at odds with democratic values such as freedom of expression, association, and assembly, access to information, and privacy enshrined in international covenants. These opposing views on information and its use shape each country's pursuit of power, influence, and legitimacy in the information environment.

INFORMATION AND POWER, INFLUENCE, AND LEGITIMACY

Exploring the relationships between information and IW objectives is essential to understanding the mechanisms of information statecraft.

Information Power

Information is foundational to state power and has long been central to international relations. Edward Hallet Carr's 1939 seminal work suggests that "power of opinion" is one of the three forms of political power.[67] Information power refers to a state's ability to affect the behavior of other states due to their superior ability to access information or to leverage the possession of certain information to make a target state comply.[68] The US has substantial information power due to both its sophisticated intelligence apparatus that collects information worldwide and its information-sharing partnerships with allies.

Not only is information is a critical element of national power; it has become a crucial battleground in strategic competition, especially as communication technology has proliferated.[69] Indeed, "other instruments of power use information as a medium to achieve their effects; information thus acts as an enabler, or even as a multiplier, for the other instruments."[70] For example, data and information play a central role in a state's ability to create wealth.[71] Some analysts have even suggested that "data capitalism" will eventually replace finance capitalism as the global economy's organizing principle.[72]

Furthermore, there is a direct connection between the health of a state's economy and its information power.[73] The growth or retraction of an economy impacts a state's ability to exploit information power. Lastly, in open societies, the trustworthiness of information is fundamental to legitimacy and power. A democratic state's power suffers when information manipulation erodes public trust in institutions.[74] Conversely, autocrats expand their power domestically by using information to control their populations and reshape beliefs about the world.[75]

Information Influence
Information has the potential to affect the perceptions and beliefs of other states positively or negatively. Truthful information exchange fosters understanding and cooperation and highlights the positive potential for information influence. Conversely, the proliferation of false information alters perceptions of common truth, which can be used to divide a society. The global penetration of the internet has revolutionized the potential for information to influence other actors.[76] For example, social media manipulation is a powerful tool for political influence, as illustrated by Russian interference in the 2016 US election.[77] It also demonstrates the expansion in scale, scope, and reach of foreign malign influence.

Russia's disinformation campaign successfully undermined public faith in the democratic process, discredited a candidate, and fueled discontent.[78] Likewise, China often uses media and information tools, including its state media outlets, social media platforms, deals with local media in other countries, and training programs for foreign journalists to influence beliefs about China and its global role.[79] Furthermore, China uses information to promote technology-enabled authoritarianism facilitated by globalization.[80]

Information Legitimacy

The essence of information legitimacy is convincing domestic and foreign audiences that a state is truthful in its messaging and its information is trustworthy. This is a challenge for autocracies, as controlling information is a core national interest and can be difficult to manage. However, technological solutions for denying information freedom, implementing censorship, and manipulating information have helped autocracies maintain such control. Conversely, "the ideas and practices of liberal democracies, including free elections, independent journalism, civic engagement, public activism, and the competition of ideas, challenge the legitimacy and power of autocrats."[81]

At the same time, democracy's reliance on legitimacy, derived from the democratic process and the outputs of government decision-making, can be fickle and delicate.[82] Public opinion is susceptible to sudden changes and manipulation, especially as the internet enables foreign and domestic actors to engage with citizens living in democracies directly. This allows adversaries to influence the legitimacy of democratic institutions and use a country's citizens against its own government.[83] Additionally, the internet and social media allow adversarial actors, both foreign and domestic, to interfere with the function of democracies by attacking consensus, questioning voting, and stoking internal conflict.[84]

SNAPSHOT OF PROMINENT US INFORMATION TOOLS

This section assesses how the US has used and misused information statecraft to achieve power, influence, and legitimacy. There are five ways in which information can be used as a tool of statecraft: information sharing, information manipulation, infrastructure destruction or sabotage, international agreements, and prosecution of individuals who use information for nefarious purposes.

Tools of Information Statecraft 153

Figure 7. Typology of Information Statecraft Tools.

Information Sharing

Tools that allow states to amplify or dispel narratives, spread awareness, and expand access to facts and information among domestic and foreign audiences

Sample Tools:
- Declassifying intelligence
- Funding credible news outlets
- Radio and TV broadcasting

Contribution to Irregular Warfare goals:
1. *Power*: Enables other IW instruments through superior info access, while denying authoritarian adversaries total control over information
2. *Influence*: Fosters greater understanding and willingness to cooperate
3. *Legitimacy*: Supports the widespread access to truth necessary for underpinning the liberal, rules-based world order

Information Manipulation

Tools that enable states to control and distort the flow of information, allowing for the distribution of false or skewed narratives. Often hurts democracies more than autocracies

Sample Tools:
- Dis- and misinformation campaigns
- Election interference
- Spreading false accounts

Contribution to Irregular Warfare goals:
1. *Power*: Affects the decision-making and behavior of adversary leaders
2. *Influence*: Alters perceptions of an adversary's (or one's own) society
3. *Legitimacy*: Erodes the legitimacy of other states' governments, but can often backfire by eroding one's own legitimacy

Infrastructure Destruction

Tools that attack or undermine an adversary's information infrastructure, such as telecommunication, transportation, elections and financial systems

Sample Tools:
- Blocking internet access
- Targeting radio transmission towers

Contribution to Irregular Warfare goals:
1. *Power*: Undermines the functioning of an adversary state, due to the criticality of information infrastructure
2. *Influence*: Shapes adversary's perceptions and understanding of the environment by denying access to information
3. *Legitimacy*: Undermines the legitimacy of authoritarian adversaries by preventing them from controlling information

International Agreements

Tools that promote multilateral and bilateral action to set and perpetuate global norms and principles for interstate interaction

Sample Agreements:
- 1976 International Covenant on Civil and Political Rights
- 2022 Declaration for the Future of the Internet

Contribution to Irregular Warfare goals:
1. *Power*: Constrains the behavior of adversaries who seek to manipulate information
2. *Influence*: Facilitates cooperation among allies and partners
3. *Legitimacy*: Delegitimizes information manipulation and promotes legitimacy of governments supporting these institutions

Prosecuting Information Criminals

Tools that involve the criminal prosecution of nefarious actors and individuals who perpetrate information operations

Sample Prosecutions:
- 2021 indictment of two Iranian nationals involved in cyber-enabled election interference

Contribution to Irregular Warfare goals:
1. *Power*: Undermines the ability of adversaries to conduct information operations abroad
2. *Influence*: Deters potential criminals from future activity
3. *Legitimacy*: Enforces laws, norms, and legal regimes pertaining to information

Information Sharing

The information environment in democracies generally lends itself to sharing information, giving democracies an advantage vis-à-vis their authoritarian competitors. Information can be shared with (or withheld from) domestic populations or foreign audiences as a tool of statecraft. For example, before the 1994 intervention in Haiti, the US broadcasted radio and TV messages from deposed President Jean-Bertrand Aristide, explaining that the US military mission was to restore democracy. The US also dropped portable radios to ensure the message was heard throughout the broader populace.[85] Information sharing facilitated a successful US intervention.

A state may leverage its relationships with allies and partners to share information about threats and collaborate on responses. The US government shares information with the global public through various media. Its involvement in radio broadcasting began in the 1940s and was codified when the Voice of America's (VOA) charter (Public Law 94-350) was signed into law.[86] The principles that guide VOA broadcasts and its subsidiary organizations[87] include serving as a reliable and authoritative news outlet; representing the whole of American society; and presenting US policies clearly and effectively. These principles continue to guide its operation today.[88] Its parent organization, the US Agency for Global Media (USAGM), works to counter misinformation and disinformation by authoritarian powers and advance American interests by providing accurate reporting to societies without free media.[89] Importantly, a legislative firewall protects the journalistic independence of USAGM (and its subsidiary organizations) and ensures they operate without allegiance to any particular political party.

The US has long depended upon various organizations and agencies for information sharing. Although Cold War–era institutions like the US Information Agency (USIA) and the interagency Active Measures Working Group[90] have been disbanded, there are now successors for the current era of strategic competition. In November 2017, the Federal Bureau of Investigation (FBI) formed a similar group, the Foreign Influence Task Force (FITF),[91] which defends against foreign interference in American democratic institutions. It focuses specifically on electoral interference and clear violations of US law.[92] Under the Department of Justice, the FITF collaborates with other departments and agencies of the intelligence community, also drawing upon the FBI's relationships with social media companies to share information about foreign interference operations.[93] Finally, the DOS-led interagency Global Engagement Center (GEC) is responsible for many of the information-sharing activities formerly under the purview of USIA, and actively shares information regarding Russian and Chinese disinformation aimed at undermining or influencing the US, its allies, or its partners.[94]

Though most US government efforts to disseminate or withhold information are directed towards foreign audiences, informing domestic audiences of foreign malign influence is essential to retaining trust in democratic institutions and protecting the notion of objective truth. To this end, declassifying intelligence to inform the domestic public is efficacious. One such example was the declassification of the Office of the Director of National Intelligence's (ODNI) assessment of the Russian influence campaign targeting the 2016 US election.[95] The ODNI report exposed the Russian campaign in detail and led to calls for action that resonated in Congress with bipartisan action and in media outlets.[96]

Information Manipulation
Governments may manipulate information to promote their own narratives or diminish those of a competitor. For democracies, this can be risky because it has the potential to erode democratic institutions and undermine trust in government.[97] When discovered, deliberate information manipulation also reduces the global legitimacy of democracy, making it more difficult for democratic governments to build and exercise soft power abroad.

Reinforcing the perception that information manipulation is pervasive adds to truth decay, which will ultimately do more harm to democratic societies than to their authoritarian competitors. There are quite a few US examples that demonstrate this point. Past instances include a false account of the Gulf of Tonkin incident to escalate the Vietnam War,[98] planting stories in both the Guatemalan domestic press and the international press to foster government instability there in the 1950s,[99] a disinformation campaign in Sweden in the 1950s and 1960s to disempower labor unions,[100] and numerous Cold War–era election interference attempts to gain influence vis-à-vis the Soviet Union.[101]

Since the Cold War era, and in particular since the Watergate scandal,[102] the US government has enacted a series of reforms to curtail official government deception and curb government power, including the Privacy Act, which restricts government use of personal data; and the Foreign Intelligence Surveillance Act (FISA), which limits the power of intelligence agencies to spy on American citizens.[103] Other important legislation includes the Federal Advisory Committee Act (FACA) of 1972 and Government in the Sunshine Act of 1976, which provides open access to public information about government surveillance when setting policy and requires publicly held meetings. The US may be using information manipulation, but given the classified nature of such efforts, evidence of recent use is not in the public domain. Regardless, American competitors employ information manipulation widely, which can facilitate gains in relative influence and legitimacy.

Destruction or Sabotage of Information Infrastructure
Destroying or sabotaging infrastructure is a tool of military statecraft that easily translates into the information domain. This is especially true given the criticality of information infrastructure (such as telecommunications, transportation, election systems, and financial systems) to the functioning of modern society.[104] In Afghanistan in 2001, one of the first steps of the US campaign was to destroy Taliban radio transmission towers—a critical means of controlling the population.[105]

More recently, the US has targeted adversary information infrastructure as part of its persistent engagement focused cyber strategy. For example, the US military's Cyber Command (CYBERCOM) blocked internet access to Russia's IRA to prevent their interference in the 2018 midterm elections.[106] The US National Security Agency and CYBERCOM conducted a similar operation against Iran as part of an effort to protect the 2020 presidential election from foreign interference.[107]

International Agreements
International agreements help set and perpetuate global norms and basic principles for interstate interaction. In response to the growing presence of divisive propaganda and disinformation campaigns during the Cold War, the UN General Assembly included an article that prohibited any propaganda for war in the International Covenant on Civil and Political Rights (ICCPR). Despite this stipulation and its adoption in 1976, the article is largely ignored in practice.[108] More recently, the UN Human Rights Council has aimed to promote and protect human rights on the internet through a series of five resolutions spanning 2012 to 2021.[109]

Another attempt to create international consensus in the information space was set forth in the 2018 Paris Call for Trust and Security in Cyberspace, which includes a commitment to strengthening capacity to prevent foreign malign influence aimed at undermining electoral processes through malicious cyber activities.[110] Though dozens of governments (including the US) and hundreds of private-sector entities and organizations have signed this nonbinding agreement, China and Russia have refused to do so.[111] In 2022, the Biden administration launched the Declaration for the Future of the Internet, subsequently signed by 61 countries, with the intention of promoting a renewal of global commitment to internet freedom.[112] These kinds of international commitments help delegitimize information manipulation as a tool of statecraft and legitimize those who agree to them.

Prosecution of Nefarious Actors

Governments may also disrupt foreign information operations by criminally charging individuals who perpetrate them. An example is the indictment of two Iranian nationals for their involvement in a cyber-enabled campaign to influence American voters in the 2020 US presidential election.[113] These individuals allegedly sent threatening email messages to intimidate voters, created and disseminated a video containing disinformation about purported election infrastructure vulnerabilities, attempted to access several states' voting-related websites, and successfully gained unauthorized access to a US media company's computer network that would have provided the conspirators another vehicle to disseminate false claims after the election.[114]

RECOMMENDATIONS

Tomes have been written suggesting means of addressing threats in the information environment.[115] Much can and should be done to address the threat that problematic information, particularly the narratives advanced by Russia and China, poses to the US. The following are recommendations for improving US information statecraft as a means of IW.

Create a Shared Understanding of Information and its Dimensions

Well-educated information consumers who can discern fact from fiction are the best defense against misinformation and disinformation. Helping audiences understand what constitutes a credible source of information and the dangers of spreading false information is essential in a world where every internet user can create content. Given the rapidity of technological change—especially the rise of generative AI and large language models as content creators—systemic education is absolutely necessary.

Information and media literacy should be incorporated into primary and secondary education to help prevent the spread of disinformation.[116]

Moreover, free and accessible continuing education related to how individuals can discern fact from fiction and recognize what counts as a credible source should become a public service. Promoting or certifying information authentication can also provide consumers with a resource to ensure what they are reading or hearing is factual.

Develop a Comprehensive National-Level Information Strategy
Foreign disinformation became a focus of the US government after Russia's state-sanctioned attempts to interfere in the 2016 election. In response, the 2017 NDAA included the Countering Foreign Propaganda and Disinformation Act, which established the GEC under the State Department.[117] Several additional organizations were created to address information-related threats. Within the Department of Homeland Security (DHS) this includes the Countering Foreign Influence Task Force (CFITF), Foreign Influence and Interference Branch (FIIB), Disinformation Governance Board, and Cybersecurity and Infrastructure Security Agency (CISA). There is also the FBI's Foreign Influence Task Force (FITF) and the Pentagon's Influence and Perception Management Office (IPMO). Finally, the Foreign Malign Influence Center (FMIC), under the DNI, the successor to the 2019-created ODNI Election Threats Executive, was established in September 2022.

A result of the rapid proliferation of multiple independent entities with overlapping missions has been incoherent execution. To rectify this, the US government should develop a unified information strategy that cuts across all aspects of the economy, society, and national security.[118] Furthermore, the private sector must be included in both strategy formulation and its execution. This whole-of-society information strategy should include clear objectives for both domestic and foreign audiences. It should also establish a lead agency or office that would be responsible for creating and implementing the strategy. Defining clear roles and responsibilities for the agencies involved in information statecraft and routinizing the interagency process for coordinating and deconflicting agency efforts are essential. Because companies and individuals can be both targets and perpetrators in disinformation campaigns,[119] the strategy must also address how to facilitate information sharing between the government, the private sector, and civil society, which is vital to defending foreign interference in the information environment.[120]

Finally, overt and covert information warfare campaigns, specifically towards Russia and China, should also be developed to support the information strategy.[121] Such campaigns must be grounded in truth to maintain the legitimacy and power of democratic norms. Whereas there are the occasional tactical and operational needs for deception and disinformation, these should always be rare exceptions. However, facilitating information access to those stuck behind digital firewalls in authoritarian states may require covert operations.

Support High-Quality Journalism at Home and Abroad
Journalists have been increasingly under attack in the US. In 2020, more than 100 journalists were arrested or criminally charged in relation to their reporting, and around 300 were physically assaulted.[122] This is a particularly disturbing trend: a free press is a cornerstone of democracy, fostering transparency, accountability, and informed decision-making. In a world where information flows through numerous media, high-quality journalism ensures that citizens have access to accurate information central to their individual and collective decision-making processes. Moreover, journalism serves as a watchdog, facilitates discourse, protects human rights, fosters transparency, and establishes agendas for debate.

Washington should recommit to supporting and protecting journalism at home and abroad, particularly in places where democracy is backsliding. Though the US government cannot, and should not, make laws restricting freedom of the press, it could restrict press access and interaction of government employees and agencies to only professional journalists belonging to organizations with a code of ethics that polices their membership, such as the Society for Professional Journalists.[123] Additionally, the US should promote freedom of information globally—not merely because it is foundational to democratic values but also because it disadvantages authoritarian governments whose power depends on strict control of information.[124] The US should continue to fund its international broadcasting assets, such as the VOA and related entities, to counter disinformation among foreign audiences.[125]

Expand Information Cooperation with Allies and Partners
Disinformation and misinformation are global threats, targeting not only the US but also its allies and partners. Sharing information about threats and evolving tactics can dramatically improve the speed and effectiveness of response. For example, Eastern European countries, particularly former Warsaw Pact states, are often the first target of Russian disinformation campaigns. Increased engagement with them could provide early warning of what Russia might attempt in the US or elsewhere. It could also provide an opportunity to build the capacity of US allies.[126]

Conclusion

The US has long prided itself on freedom of speech and information; in many ways, these principles have afforded it power, influence, and legitimacy in the international system. Information statecraft is essential to the health of its domestic polity and in prevailing against competitors. As autocracies increasingly weaponize information against democratic institutions, the US must strategize and organize quickly to counter adversary narratives. The importance of an educated public that understands and recognizes false information is essential to strategic competition, in which ideas and data are the prevailing currency.

Notes

1. Rosenbach and Mansted, "The Geopolitics of Information," 1.
2. Wong, "The Age of Informational Statecraft."
3. Wardle and Derakhshan, "Information Disorder."
4. Brandt, "An Information Strategy for the United States."
5. Howard et al., "The IRA."
6. Brandt, "How Autocrats Manipulate."
7. Filloux, "You can't sell news"; and ," 39.
8. Yablovkov, "Russian Disinformation."
9. Wardle and Derakhshan, "Information Disorder."
10. Prier, "Commanding the Trend," 51.
11. Howard et al., "The IRA," 39.
12. Mejias and Vokuev, "Disinformation and the Media."
13. Saressalo and Huhtinen, "Information Influence Operations"; and Office of the Director of National Intelligence, *Assessing Russian Acitivies and Intentions in Recent US Elections.*
14. Howard et al., "The IRA, Social Media and Political Polarization," 4.
15. Howard et al., 8.
16. Howard et al., 17.
17. World Health Organization, "Infodemic."
18. Bandeira et al., "Weaponized."
19. Lisedunetwork, "Information: Types of information."
20. Joint Chiefs of Staff, *Information in joint operations* (JP 3-04), 2022.
21. Murphy and Kuehl, "The Case for a National Information Strategy"; and Finley, "A 3Cs Framework."
22. Cyberspace Solarium Commission, "Countering Disinformation."
23. Theohary, "Information Warfare."
24. Mull and Wallin, "Propaganda."
25. Theohary, "Information Warfare."
26. Bennett and Livingston, "The Disinformation Order."
27. Erlich and Garner, "Is pro-Kremlin Disinformation Effective?"
28. Foreign Malign Influence Center, 50 USC. § 3059 (e)(2), (2022), 50 USC § 3059(e)(2)
29. Kavanagh and Rich, *Truth Decay.*
30. Kavanagh and Rich.
31. Gris et al., *Rivalry in the Information Sphere.*

32. Wong, "The Age of Informational Statecraft,"
33. Larson et al., *Foundations of Effective Influence Operations.*
34. Theohary, "Information Warfare," 1.
35. Theohary, 2.
36. White House, *2017 National Security Strategy.*
37. Wong, "The Age of Informational Statecraft."
38. Wanless and Pamment, "How Do You Define a Problem?"
39. Larson et al., *Foundations of Effective Influence Operations.*
40. Wanless and Pamment, "How Do You Define a Problem?"
41. Sun Tzu, *The Art of War.*
42. Curtis, "An Overview of Psychological Operations (PSYOP)."
43. Curtis; and Gopnik, "How Samuel Adams Helped."
44. Mariarosaria Taddeo, "Just Information Warfare."
45. Porche et al., "Redefining Information Warfare Boundaries."
46. Wood, "Information Engineering."
47. Prier, "Commanding the Trend."
48. Porche et al., "Redefining Information Warfare Boundaries."
49. Theohary, "Information Warfare."
50. Theohary.
51. Prier, "Commanding the Trend" ; and Wanless and Pamment, "How Do You Define a Problem Like Influence?"
52. Joint Publication, *Information Operations.*
53. Sherman, "Untangling the Russian Web"; and Jacobs and Carley, "#WhatIsDemocracy."
54. Brandt et al., "Linking Values and Strategy," 4.
55. Brandt et al.
56. Nye and Owens, "America's Information Edge."
57. Nye and Owens.
58. Arquilla and Ronfeldt, *The Emergence of Noopolitik.*
59. Arquilla and Ronfeldt.
60. The White House, "International Strategy for Cyberspace."
61. The White House.
62. Brandt et al., "Linking Values and Strategy," 7.
63. Guriev and Treisman, *Spin Dictators*, 9.
64. Hall and Ambrosio, "Authoritarian learning."
65. Hicks et al. "By Other Means Part I."
66. Brandt et al., "Linking Values and Strategy," 8; and Shapiro and Wanless, "Why Are Authoritarians Framing?"
67. Carr, *The Twenty Years' Crisis, 1919–1939.*

68. Raven, Schwarzwald, and Koslowsky, "Conceptualizing and Measuring."
69. Saressalo and Huhtinen, "Information Influence Operations."
70. Saressalo and Huhtinen, 41.
71. Rosenbach and Mansted, "The Geopolitics of Information," 3.
72. Mayer-Schönberger and Ramge, *Reinventing Capitalism*.
73. Worley, *Orchestrating the Instruments of Power*, 243.
74. Kuo and Marwick, "Critical disinformation studies."
75. Guriev and Treisman, *Spin Dictators*, 4.
76. Rosenbach and Mansted, "The Geopolitics of Information," 2.
77. Bradshaw and Howard, "The Global Organization."
78. Klein, "Information Warfare."
79. Kumar, "How China uses the news media."
80. Kurlantzick, *Beijing's Global Media Offensive*, 6; and Economy, *The World According to China*.
81. Brandt et al., "Linking Values and Strategy," 7.
82. Reisher, "The Effect of Disinformation," 45.
83. Reisher.
84. Baer-Bader, "EU Response to Disinformation."
85. Singer, "Winning the War of Words."
86. Heil, *Voice of America*; and Voice of America, "VOA's Mission in the 1960s and 1970s."
87. These same principles guide Radio Free Europe, Radio Liberty, Office of Cuba Broadcasting, Radio Free Asia, Middle East Broadcasting Networks (which include TV and radio).
88. US Agency for Global Media, "Mission."
89. US Agency for Global Media, "Truth over Disinformation."
90. Jones, "Going on the Offensive."
91. Bodine-Baron et al., *Countering Russian Social Media Influence*.
92. US Department of Justice, "Deputy Assistant Attorney General Adam S. Hickey."
93. FBI, "Combating Foreign Influence."
94. US Department of State, "Mission & Vision."
95. Office of the Director of National Intelligence, *Assessing Russian Activities*.
96. Gazis, "Bipartisan Senate Intel Report."
97. Brandt, "An Information Strategy for the United States."
98. Paterson, "The Truth About Tonkin."
99. Grandin, "Off the Beach."
100. Nilsson, "American Propaganda, Swedish Labor."

101. Ward, Pierson, and Beyer, "Formative Battles."
102. Berger and Tausanovitch, "Lessons From Watergate."
103. Foreign Intelligence Surveillance Act of 1978.
104. Wilson, "Cyber Threats to Critical Information Infrastructure."
105. Singer, "Winning the War of Words."
106. Nakashima, "U.S. Cyber Command operation."
107. Cohen, "US recently conducted cyber operation."
108. Kearney, "The Prohibition of Propaganda for War."
109. Article 19, "UN: Human Rights Council"; UN General Assembly, "The promotion, protection, and enjoyment of human rights."
110. Paris Call For trust and security in cyberspace, "Home."
111. Lété, "The Paris Call."
112. White House, "A Declaration for the Future of the internet."
113. US Department of Justice, "Two Iranian Nationals."
114. US Department of Justice.
115. Helmus and Kepe, "A Compendium."
116. Yablokov, "Russian Disinformation."
117. Hall, "The New Voice of America."
118. Office of Inspector General, *DHS Needs a Unified Strategy*; and Rosenbach and Mansted, "The Geopolitics of Information," 16.
119. Worley, *Orchestrating the Instruments of Power*, 434.
120. Brandt et al., "Linking Values and Strategy," 21.
121. Seth G. Jones, "The Return of Political Warfare."
122. Jacobsen, "In 2020, US Journalists Faced Unprecedented Attacks."
123. Society of Professional Journalists, "Home."
124. Brandt, "An information strategy."
125. Curtis, "Springing the 'Tacitus Trap.'"
126. Moy and Gradon, "COVID-19 Effects and Russian Disinformation," 17.

Chapter 8

Tools of Resilience

> Applying a resilience lens to great-power competition will require the United States to systemically identify and prioritize critical assets, capabilities, functions, and dependencies. It means asking (and answering) difficult questions: Where is America most vulnerable, and how critical are these vulnerabilities? What functions are essential and must be prioritized? And conversely, where can the United States assume some risk? [1]
>
> Erica D. Borghard

Introduction

Entering the word "resilience" into the search field of Amazon's bookstore results in a list that covers everything—from children's coloring books designed to build grit to cybersecurity and civil engineering texts. Indeed, resilience has many discipline-dependent definitions, and national resilience is an aggregate concept. Resilience is hydra-headed and, as such, is difficult to analyze holistically. Nonetheless, it is a critical factor in navigating evolving geopolitical competition and thriving in an era of crises. Truth decay will make the information environment key terrain in the IW realm. Robust cybersecurity will be essential for ensuring governmental, economic, and social resilience. Additionally, principles from special operations' FID doctrine and concepts guiding indigenous resistance to authoritarianism are valuable resources when considering ways to bolster resilience at the national level.[2]

When considering resilience through the lenses of national security and IW, one must evaluate the implications of climate change on military readiness, humanitarian disasters, and resource wars.[3] Doing so requires using analytical tools from many disciplines, including public health, engineering, sociology, cybersecurity, and political science. The importance of leveraging all disciplines is apparent when examining the costs of such crises. In seventy-five years, federal revenue losses could exceed $2 trillion annually, and the US government could spend an additional $125 billion annually on expenses such as disaster relief, disaster insurance, fire suppression, and flooding at federal facilities.[4] These estimates do not include expenditures for international crises.

Given its manifold impacts on national security, resilience should be a central pillar of contemporary American grand strategy.[5] Yet despite two Presidential Policy Directives (PPD-8 and PPD-21)[6] defining and prioritizing national resilience, the inclusion of a resilience-dedicated section in the 2017 NSS,[7] and resilience being mentioned twenty-six times in the 2022 NSS,[8] the US has largely failed to connect resilience and its requirements to national security in a meaningful way. Developing a strategy for increasing and assuring national resilience will require taking calculated risks and making difficult tradeoffs.[9]

This chapter defines resilience and describes its core attributes before illustrating how resilience relates to IW and contributes to its objectives. It then articulates four levels of analysis—individual, community, national, and international—explaining the tools available at each level for building resilience. Finally, the chapter recommends operationalizing the nascent concept of collective resilience as a foundation for prevailing in an age of strategic competition and enduring crises.

Defining Resilience

The word "resilience" derives from the Latin *resilio*, which means to jump back or rebound.[10] However, many contemporary definitions are far broader, partially due to disciplinary differences. Andrew Zolli and Ann Marie Healy's *Resilience: Why Things Bounce Back* is a touchstone for many works concerning national resilience.[11] Their central proposition contends that "a large piece of resilience is about preserving adaptive capacity," which they define as "the ability to adapt to changed circumstances while fulfilling one's core purpose."[12] We adopt this definition and view resilience in the broadest terms. A lack of resilience leaves a state vulnerable to the active IW efforts of competitors.

Since several key characteristics of resilience have not been fully resolved in the literature, certain assumptions have to be made. First, resilience must be both anticipatory and reactive—if not, it is useless for strategy and policymaking. Second, resilience must be both tactically and strategically focused. There are two broad approaches to improving resilience in the literature—adapting quickly or long-term transformation,[13] both of which are valid. Third, resilience encompasses robustness and fragility within it. Robustness is the ability to withstand unforeseen shocks, whereas fragility is its opposite. Although these terms have been treated as separate variables,[14] they are included in our definition of resilience. Indeed, shocks (e.g., natural disasters) that may seem unrelated to IW can create tremendous vulnerabilities to IW by competitors when resilience is inadequate. Similarly, while most resources are allocated to natural disasters, these events are only indirectly connected to IW despite their crucial role in national resilience

Resilience and Power, Influence, and Legitimacy

Despite the importance of resilience in defending against transnational threats (e.g., climate change, pandemics, etc.), the concept is only tangentially associated with national security. Yet, resilience is intimately tied to all components of national security and has implications for an era of both crises and strategic competition. Therefore, resilience must be the fourth pillar of IW strategy in pursuing power, influence, and legitimacy. The role of resilience in national security is twofold. First, there is the role resilience plays in a state's power, influence, and legitimacy. For example, all states in the international system must contend with crises, and their effectiveness in doing so directly contributes to their performance in strategic competition. This was evident during the COVID-19 pandemic. China's Zero-COVID policy and vaccine diplomacy were intended to showcase the resilience of its society and the effectiveness of its governance model. But these draconian policies backfired, diminishing China's global influence and legitimacy.[15] Thus, resilience can directly impact a state's power, influence, and legitimacy, along with other IW tools.

Second, resilience in the form of preventative measures plays an important role in national security; resilience matters just as much for what it prevents as for what it achieves. A negative aim seeks to deny the adversary from achieving its own positive aim, whereas a positive aim entails the achievement of a specific objective. This logic applies directly to IW. China and Russia are adept at wielding a variety of IW instruments—from paramilitary forces to economic coercion and disinformation campaigns—to bolster their power, influence, and legitimacy at the expense of others. Resilience embodies a means to defend against those activities. If military, economic, and information tools help the US achieve positive aims, then resilience tools can pursue negative aims—namely, the defense of its own power, influence, and legitimacy, thereby denying adversary IW campaigns. Given the simultaneously positive and negative role of resilience in IW, the distinct links between resilience and each of the three IW objectives of power, influence, and legitimacy will be addressed next.

Power and Resilience
The COVID-19 pandemic was humbling for some governments more than others. Despite its impressive military arsenal and economic strength,

the US lagged behind in its ability to contain and manage the pandemic.[16] Therefore, the pandemic's impacts demonstrated the need to expand conceptions of power in an age of cascading crises and to appreciate the danger of hidden vulnerabilities in traditional conceptions of power. Seemingly powerful states can topple with little warning. Indeed, over the last thirty years, "the sudden collapse of the Soviet Union and the breakdown of state authority in the aftermath of the Arab Spring suggest states can be fragile in ways that conventional measures of power do not capture."[17]

The National Intelligence Council's 2017 Global Trends report argues that "measuring a state's resilience is likely to be a better determinant of success in coping with future chaos and disruption than traditional measures of material power alone."[18] Traditional calculations of state power, such as military spending, population size, and GDP, fail to portray the degree to which a state is resilient (or not).[19] Political-psychological aspects of resilience has been deemed important, too, in calculating state power.[20] Therefore, traditional calculations of power should include calculations of resilience.

Giske concludes that in an era of crises, resilient states are better able to both deter and recover from shocks.[21] In "The Pandemic is Revealing a New Form of Power," Uri Friedman observes that COVID-19 pandemic "revealed the salience of resilient power."[22] Such power relies on domestic attributes like quality and scale of critical infrastructure, confidence in government, and strong alliances. Simply put, there is a direct relationship between resilience and power; states that can withstand, adapt, and recover from crises, manmade or natural, are more powerful. This provides a key advantage in strategic competition, as the states that better mitigate and recover from crises will have more readily available power, which they can wield against their adversaries.

The relationship between resilience and defense is also important to a state's power, as Michael T. Klare links emerging crises to an inevitable erosion in military power. For example, flooding and high winds have already destroyed portions of US military installations.[23] Klare further suggests that "in the worst case scenarios, the major powers will fight over power and other resources, producing new global rifts."[24] Thus, resilience and power are connected in two distinct ways. First, an inability to harden critical infrastructure will degrade state capability and power. Despite common biases toward military hardware (e.g., planes and ships), critical infrastructure like resilient energy grids and water supplies is just as important for national power. Second, extreme weather events or other shocks can trigger greater interstate competition for scarce resources, creating vulnerabilities to IW approaches.[25] Impending crises, how they are managed, and the degree to which a society is able to rebound from them will directly impact state power and, by extension, the balance of power in the international system. Resilience is thus directly related to strategic competition and IW.

Influence and Resilience
There is an interdependent relationship between a state's level of resilience and its influence, both domestically and internationally. The ability to change public sentiment brings the potential "to influence policy, disrupt economics, incite violence, and even provoke civil war."[26] Influence and resilience both rely on trust, particularly on accurate and truthful information.

One of the principal goals of contemporary IW is to erode trust among adversary populations. Indeed, the rise of foreign interference in political processes is fundamentally aimed at destroying trust and manipulating public opinion.[27] David H. Ucko and Thomas A. Marks suggest that states "seek to accentuate (or mitigate) the schisms in targeted societies, thereby to gain power and influence."[28] The *Kremlin Playbook* describes the combination of activities that Russia has used to gain access and influence, aiming at "weakening the internal cohesion of societies and strengthening the perception of the dysfunction of the Western democratic and economic system."[29] Greater resilience in the face of these IW campaigns would prevent adversaries from increasing their own influence and decreasing that of the US.

Internationally, the trust in partners to uphold their commitments is foundational to informal and formal alliances, which add to national resilience. Heather J. Williams and Caitlin McCullough note that "though experts debate the degree to which credibility is important in maintaining alliances, they agree it is a relevant factor."[30] Credibility is similarly important to domestic audiences, as illustrated in the varied responses to the COVID-19 pandemic and their relative effectiveness. A country's ability to deal with the pandemic did not depend on regime type; instead, the crucial determinant of a state's performance was a combination of its capacity and trust in government.[31] A principal reason the US lagged behind Australia in its pandemic response was that Americans did not trust their government to tell the truth as much as Australians trusted their elected leaders.[32] Resilience levels covary with levels of political trust. Less trust means lower levels of resilience[33] and increased susceptibility to IW campaigns.

Legitimacy and Resilience

Legitimacy, the belief that something is right or appropriate, is a vital source of resilience. In fact, organizational survival and success depend on legitimacy.[34] This belief is socially constructed; legitimation is a process in which shared cultural beliefs create expectations for what is likely or should occur in a particular situation.[35] For example, in some cultures, community needs take precedence over individual ones. Likewise, in some states, stability is valued over freedom. These notions color the degree to which citizens hold certain expectations about who is responsible for what in a society. As it pertains to resilience, this means that citizens believe that their government has put in place the appropriate policies to prepare for crises, from reacting to and recovering from them.

Therefore, the resilience of domestic institutions is directly tied to the degree to which its citizens believe them to be legitimate. Pol Bargués and Pol Morillas suggest that government legitimacy depends on the degree to which the population perceives its institutions as efficient and able to address its needs.[36] Conversely, lack of trust in authorities, the belief that the government will not "make the right decision during times of crisis"[37] erodes legitimacy and, thus, resilience.

Trust in political institutions and processes, in particular electoral participation, is a key indicator of the legitimacy of democracy.[38] Similarly, broad population agreement regarding national security policy contributes to the ability of citizens to absorb adversity.[39] Psychological-political elements such as "strength of democracy and trust in leadership" directly contribute to national resilience.[40] The degree to which individuals, communities, or states are prepared for, able to respond to, and bounce back from crises effectively depends on what people view as the legitimate role of government institutions, organizations, and individuals in a society. For states under threat of military invasion, government policies and legal frameworks reinforce the legitimacy of post-invasion resistance efforts.[41]

Legitimacy also matters for authoritarian resilience.[42] Kimly Ngoun suggests, "authoritarian legitimation tends to rest on multiple claims including identity, democratic procedures, performance, personalism, and international engagement."[43] When government institutions fail to deliver on public goods, citizens may rely on community resources or individual connections to meet their needs, thus eroding legitimacy and resilience.[44] Regardless of government type, legitimacy and resilience are directly related—and states that lack legitimacy offer a prime target for IW campaigns that exploit societal fissures.

Tools of Resilience   175

Figure 8. Typology of Resilience Tools.

Individual Resilience

Tools cultivate individual physical, mental, and psychological resilience, building hardiness and grit.

Sample Tools:
- Self Assessment
- Individual wellness resources
- Material resources
- Accurate and timely information
- Information countermeasures
- Individual readiness

Contribution to IW goals:
1. **Power**: Increases individual capability and readiness
2. **Influence**: Cultivates greater awareness of strengths and weaknesses
3. **Legitimacy**: Improves perceptions of ability to cope

Community Resilience

Tools strengthen community infrastructure, social and economic well-being, connectedness and cohesion.

Sample Tools:
- Emergency planning
- Assessments
- Guides and resources
- Data/Modeling
- Public-private partnerships
- Cooperative norms

Contribution to IW goals:
1. **Power**: Empowers communities to withstand and recover from crisis
2. **Influence**: Reassures individual and communities of readiness
3. **Legitimacy**: Builds connection and cohesion for appropriate crisis response

National Resilience

Tools allow states to prepare for, respond to, and bounce back from shocks with core values and institutions intact.

Sample Tools:
- Critical infrastructure protection and public goods
- Diversified supply chains
- Transparent, truthful and timely information
- National Protection Plans.
- Continuity of government

Contribution to IW goals:
1. **Power**: Deters manmade crises
2. **Influence**: Bolsters confidence in government
3. **Legitimacy**: Improves trust in institutions

International Resilience

Tools contribute to resilience of the international system and interactions of states, alliances, international organizations, and multinational corporations.

Sample Tools:
- Resource int'l organizations
- Diversified and insulated supply chains
- Robust formal alliances
- Strong bilateral relationships
- Collective resources for crisis response

Contribution to IW goals:
1. **Power**: Builds stronger partners to withstand crises
2. **Influence**: Erodes alliances and partnerships of adversaries
3. **Legitimacy**: Strengthens international system resilience and undermines adversary legitimacy

Tools of Resilience

This section examines the factors underpinning resilience and their implications for the US and its allies and partners. It introduces four levels of resilience—individual, community, national, and international—and explains the available tools. Although distinct, the four levels are interdependent; therefore, they must be managed in concert. The tools of resilience, at each level, provide opportunities to gain relative power, influence, and legitimacy.

Individual Resilience

Individual resilience comprises "a person's capacity for adapting psychologically, emotionally, and physically...without lasting detriment to self, relationships or personal development in the face of adversity, threat or challenge."[45] George A. Bonnano frames individual resilience in terms of hardiness, which consists of three dimensions: "being committed to finding meaningful purpose in life, the belief that one has some influence over one's surroundings and the outcomes of events, and the belief that one can learn from both positive and negative experiences."[46]

In addition to the psychological and emotional components, like hardiness, material factors also matter. Access to a variety of resources is fundamental to preparation for shocks and crises. For example, factors like income, education, and social support are predictors of resilience.[47] The Fund for Peace's State Resilience Index (SRI) includes individual capabilities such as "a stock of education, health, income, and food security...so that when a crisis hits, people are not immediately rendered destitute"[48] as important contributors to resilience. Thus, the overall welfare of individuals and the amount they each have in reserve (both psychologically and materially) are crucial factors contributing to individual resilience.

Self-diagnosis of one's strengths and vulnerabilities can provide a foundation for building resilience. The Resilience Scale[49] and the Brief Resilience Scale[50] are free, publicly accessible tools that individuals can use to diagnose their ability to bounce back. Understanding one's own capacity to withstand shocks can be an important first step to improving it.

Activities such as building knowledge and skills, developing positive mental models, creating community connections, and belonging to social groups not only improve individual wellness but also contribute to personal resilience.[51] Though stressful situations negatively impact individual resilience, proper nutrition, ample sleep, adequate hydration, and regular exercise can reduce their toll.[52] Likewise, mindfulness can help individuals build connections and restore hope, improving resilience.

In a crisis, timely, accurate, and truthful information can bolster individual resilience.[53] Given the prevalence of misinformation and disinformation, developing individually focused countermeasures is increasingly important. In addition to education and access to fact-checking resources, government transparency and data sharing can also be useful for building resilience.[54] Inoculation theory—applying the principles of medical vaccination to the realm of persuasion—as a tool for countering disinformation and misinformation suggests that if people are provided with facts and arguments preemptively (as protection against disinformation and misinformation), they become more resistant to it.[55]

Clearly delineating individual roles and responsibilities during times of crises, as well as setting expectations about the community and national level support available, can bolster individual resilience. For example, Finland imposes an expectation on its citizens that households should be able to survive for seventy-two hours without state assistance during a crisis.[56] Instead of relying on norms or experience with past shocks to drive present or future behavior, public and private actors can play a critical role in setting expectations about what sort of government support individuals can expect in a time of crisis. This can incentivize better individual preparedness, reaction to shocks, and recovery.

Finally, providing material support to prepare for and recover from crises contributes to individual resilience. This can include free or subsidized access to vaccines and testing kits in the event of pandemic disease or sandbags and plywood during inclement weather. In certain instances, state or national authorities may decide such material support is required. The state's support for building individual resilience also enhances its legitimacy and contributes to community and social resilience.

Community (and Social) Resilience
Regardless of a country's size, disasters occur locally, requiring local governments to act as first responders, with support from higher levels of government following later. In the US, the DHS assigns responsibility to the lowest jurisdictional level present within the impacted area.[57] Community preparation is imperative.

The conditions under which communities exist—including their resources, social networks,[58] socially cohesion,[59] and inclusiveness [60] are fundamental factors enabling their resilience.[61] Similar to individual resilience, community resilience depends on a combination of factors, including social and economic well-being, connectivity for resource exchange, and the engagement of community organizations in crisis planning, response, and recovery.[62] Community infrastructure, comprising services and facilities, such as health services and community centers, supports and sustains community resilience.[63] Social resilience, a component of community resilience, refers to the collective ability of groups to cope with unexpected events[64] and their capacity to adapt and transform in response to them.[65] Population wellness,[66] encompassing high levels of health, functioning, and quality of life,[67] also contributes to social and therefore community resilience.

Similar to individual resilience, the first step in building community resilience is identifying strengths and weaknesses before an incident occurs. Once identified, it is crucial to develop standard operating procedures and assign responsibilities for actions in the event of a crisis. Communities have various assessment options available to accomplish these tasks.

Tools of Resilience

A recent National Institute of Standards and Technology (NIST) report inventoried fifty-six existing community resilience frameworks and tools in use in the US.[68] One example is the Communities Advancing Resilience Toolkit (CART), which includes an assessment survey for gathering baseline community information.[69] It can be updated before or after a disaster to facilitate community planning for building resilience. Using the CART requires community input and deliberation:[70] it provides a valuable outcome, and the process of gathering the information contributes to community resilience through increasing connection and improving awareness. The tools the NIST surveys vary in focus, provider, and hazard addressed,[71] and they have identified two other types of tools for developing community resilience: guides and data/modeling.[72]

Besides assessment tools, guides, and modeling, emergency planning is another way to build community resilience. Emergency planning includes assigning particular responsibilities to various organizations and institutions, from education, health, and local government to a state's armed forces.[73] During a crisis, ensuring that physical needs such as water and food are met is vital, as is providing security; the availability of community resources reinforces social cohesion and trust in leadership.[74]

The economic well-being component of community resilience relies, to some extent, on supply chains and critical infrastructure.[75] Building redundancy for continuity during emergencies may involve developing contractual agreements with the private sector to address shortfalls before crises occur.[76] Several successful public-private partnerships have improved disaster response and recovery, such as those during the Deepwater Horizon oil rig explosion and subsequent spill, as well as the 2011 tornado in Joplin, Missouri.[77]

Finally, fostering cooperative norms among individuals and communities can cultivate optimism and cohesion during a crisis. Individual, social, and community resilience are mutually reinforcing. Planning for a crisis enhances social connections and promotes expectations of mutual support within a community, producing a virtuous cycle that enables individuals and communities to weather crises more resiliently.

National Resilience

The US government defines national resilience as "the ability to prepare for and adapt to changing conditions and withstand and recover rapidly from disruptions;"[78] and its population must also believe it has the capacity to do so.[79] The ability to withstand diverse adversities and crises includes preserving a society's core values and institutions.[80] Social factors that underpin national resilience include patriotism, optimism, social integration, and trust in political and public institutions (including national leaders and decision-makers).[81] Key national resilience components include the sustainability of government institutions; continuation of societal norms (individual liberties and civil rights in democracies); economic durability; and the stability of political, diplomatic, or military commitments abroad.[82] These ideas are well captured in NATO requirements for resilience capabilities of its member-states, which include continuity of government, provision of essential services (energy, food, water, health, communications, and transportation), and civil support to the military during crises.[83]

Assessing national capabilities and risks is a tool for improving national resilience. Assessment includes

> identifying and prioritizing critical assets, systems, functions, and personnel, and mapping the dependencies between them; understanding the threat environment and ascertaining likely avenues and methods of attack; being willing to assume greater risk in areas that may be more exposed but are of less significance, or are less likely to be targeted; and building the organizational maturity to continuously update and sustain these practices based on new information and evolving understandings of the threat and vulnerability landscape.[84]

Creating contingency plans, ensuring the continuity of government, and delivering the population's needs during shocks are essential for resilience. This includes resourcing national protection plans and, for states under threat of invasion, resistance frameworks and expectations.[85] Other tools that fall under this umbrella include protecting elections and voting systems[86] and updating, protecting, and ensuring access to critical infrastructure.[87] Given the prevalence of privately owned infrastructure and the central economic role of goods and services, enabling the private sector and securing and diversifying supply chains are essential.[88] Resilient supply chains are vital to withstanding security, economic, reputational, or workforce shocks.[89]

National resilience is improved when individual and community resilience are appropriately resourced, minimizing disruption in people's lives.[90] This includes the provision of public goods and resources, such as essential health supplies and services before, during, and after crises, as well as education, which can bolster individual resilience through building problem-solving skills.[91] Transparent, truthful, and timely government messaging is another critical aspect of national resilience. Trust is built and ensured by accurate, simple, and well-timed communication.[92] As mentioned in chapter 7, objective reporting and access to fact-checking resources are essential components for building and maintaining trust in government and its institutions. A government is not resilient if people do not believe in it.[93]

International Resilience
Support from allies, partners, and international organizations can bolster national resilience. Indeed, some suggest "modern-day resilience may require a regional approach or multinational cooperation, and will almost certainly need to consider the requirements of the private sector."[94] Unlike other levels of resilience, the global interconnectedness of states can be a double-edged sword—it can be both a source of strength and a point of vulnerability. The impacts of conflicts, mass migration, economic recession, pandemics, and climate change, as well as the effects of

globalization and new technology, all illustrate potential negative or positive consequences.[95]

NATO's conceptualization of national resilience encompasses alliance members' individual and collective capacity to withstand, navigate, and swiftly recover from disruptions caused by military and non-military threats to Euro-Atlantic security from authoritarian actors, strategic competitors, and global challenges.[96] The war in Ukraine demonstrates NATO's ability to support and reinforce a nation in crisis without direct military intervention.[97] In some ways, NATO's expertise and commitment to resilience stems from its experience defending against the Soviet Union.[98] During the Cold War, resilience meant harmonizing military capacity, civil preparedness, and emergency planning.[99] This suggests that resilience can have a deterrent effect, "requiring a state to analyze the fragility of an opposing state before taking military action."[100]

The idea of resilience as deterrence underpins the 2022 NDS, which focuses on resilience and collaboration with allies to deter aggression, promoting the idea that resilience is at the core of collective security and defense.[101] Other authors suggest that deterrence by resilience is possible.[102] But, for resilience to be a credible form of deterrence, states must be committed to scaling up resilience as a national and international priority and capable of withstanding shocks and recovering from them intact. They must also communicate their relative resilience to both domestic and external audiences and be willing to resist aggression.[103]

States have several ways to contribute to international resilience. First, states should bolster their national resilience to reduce the impacts of shocks. NATO is encouraging such an approach, asking members to monitor and evaluate state-level policies and encourage domestic efforts to bolster resilience, including securing and diversifying supply chains, protecting infrastructure, addressing the impact of emerging technologies, and securing communications systems.[104] By increasing self-sufficiency, there may be more available resources for supporting partners or allies. Second, by improving national resilience, adversary IW campaigns are more costly to prosecute. China has taken steps to build greater self-sufficiency due to its concerns that infiltration of Western values and ideology could be destabilizing.[105] It has also worked to secure alternatives to natural resources, components, and critical technologies,[106] to protect itself from American IW tools.

The COVID-19 pandemic underscored the importance of global supply chains for international resilience. In response, states have made broad efforts to diversify and insulate these chains against both manmade or natural disasters.[107] "Friendshoring" or reshoring supply chains may guard against weather- or pandemic disease-related crises and overreliance on strategic competitors.[108]

Crises that emerge from interdependence, whether they are health-, crime-, or military-related, necessitate international solutions. Strengthening ties and alliances among like-minded states to foster collective resilience can serve as a deterrent against various threats.[109] For example, Yongtao Gui suggests the US could leverage the Indo-Pacific Economic Framework to create and implement a collective resilience strategy to counter Chinese economic coercion.[110] NATO recently created rapid reaction teams to fight cyberattacks against its members.[111]

The creation and maintenance of organizations that underpin the international system were meant to build resilience transnationally. The UN and its subsidiaries, as well as organizations such as the World Bank, the WTO, and the International Red Cross, aim to insulate states from risk and support them in times of crisis. Therefore, international resilience depends on the sustainability of global systems and interactions between states, alliances, international organizations, and multinational corporations. US influence and legitimacy matter in these organizations, as does adequate funding. For example, the UN Office of Disaster Risk Reduction relies on voluntary contributions to accomplish its mission of convening partners and coordinating activities to create safer, more resilient communities.[112] Supporting global economic development activities through international organizations can also help address national-level vulnerabilities, such as public health capacity, which is essential given how easily disease travels.[113]

Recommendations

The recommendations that follow are based on the analysis of the strengths and weaknesses of the current US prioritization of resilience. Given the centrality of resilience to national security, these recommendations represent the first steps for building a grand strategy around the concept of resilience.

Prioritize and Resource Resilience
Although military readiness is crucial, the challenges of the twenty-first century necessitate a strong emphasis on national resilience. A balanced approach, recognizing the interconnectedness of these domains, will best serve national security. There will likely not be enough money to fund all priorities. Therefore, the US must assess the existing vulnerabilities and decide which capabilities to trade to address them. An emphasis on spending that addresses non-traditional threats, such as natural disasters, pandemics, climate change impacts, cyber-attacks, and infrastructure failures, can reduce vulnerability to IW approaches. This may require spending less on military hardware and more on resilient infrastructure.

Monitor National Resilience
America's allies recognize the importance of resilient power in the coming decades and are working to operationalize the pillars and frameworks they have created. The US must not only adopt a coherent, exportable, and agile framework for measuring national resilience (which will be covered in the next chapter); it must also closely monitor it. Continuously monitoring resilience helps incorporate new knowledge into policy, in addition to facilitating the institutional reforms necessary to build resilience.[114] The stakes are tremendous. For example, the predictable impacts of climate change are extraordinarily costly—estimates of weather impacts alone (in

the absence of infrastructure improvements) are expected to cost the US up to 7% of its GDP.[115] For many potential crises—such as weather-related shocks and pandemics—preventative steps are well known but rarely undertaken. Monitoring the readiness of communities can incentivize proactive measures, which not only increases national resilience, but ultimately results in cost savings.

Make Resilience Central to US National Security
Given the stakes, as noted by Erica D. Borghard, the next administration should develop its National Security Strategy based on a critical pillar of resilience.[116] The recent spate of hurricanes in Florida and wildfires on Maui are merely a preview of coming disruptions, underscoring the immediate need for improving resilience at all levels. A first step could be examining the lessons of the COVID-19 pandemic.[117]

These efforts will be expensive and politically challenging but far less than the price of inaction. Whereas robust financial, political, and analytical investments in resilience will be high, these investments are less costly than the catastrophic consequences of a systemic shock. For example, in 2020, the World Health Organization estimated that investing in reducing disease transmission would cost an estimated $30 billion per year, whereas the estimated global cost of the COVID-19 pandemic was expected to be nearly $16 trillion—roughly 500 times costlier than investing in preventive measures.[118] By the end of 2023, the cost of COVID-19 response and recovery to the US alone is estimated to reach $14 trillion.[119] The pandemic's total economic impact is twenty times greater than that of the 2001 World Trade Center attacks and forty times greater than that of any natural disaster the US has experienced in this century.[120]

Build Collective Resilience
The US should identify potential areas of cooperation with partners and allies to build national and collective resilience. NATO introduced the concept of collective resilience in 2017 as "the collective capacity of allied states to withstand, fight through, and quickly recover from disruption

caused by military and non-military threats caused by adversaries and global challenges."[121] Based on a similar foundation to collective defense, collective resilience rests on an assumption of synergistic effects: when partners and allies work in common cause, the outcome will exceed the sum of the individual parts. This idea should be conceptualized as a twenty-first-century Marshall Plan. By increasing collective resilience among like-minded states to withstand both naturally occurring crises and the effects of adversarial IW campaigns, states will have greater power, influence, and legitimacy to prevail in the ongoing strategic competition while inoculating themselves against the negative impacts of crises.

Historically, the US has been quick to respond in support of states experiencing crises, providing resource shortfalls and technical assistance. Expanding this type of cooperation before shocks can diminish the resources required and the need for immediate support in their wake. Additionally, the US should prioritize assisting allies and partners experiencing IW campaigns, in particular those geographically adjacent to competitors. This could include capacity building to deter adversaries, prepare resistance, and, if necessary, counter the military aggression of competitors. The US can augment these approaches by overtly pledging to support both violent and non-violent resistance in advance of aggression and occupation.

Conclusion

Most tools of resilience are domestic and thus have traditionally existed outside of the national security sphere. But today, these tools are critical for defending and advancing power, influence, and legitimacy relative to adversaries. The pandemic revealed the limitations of hard power, and the states likely to prevail in IW are those capable of withstanding and mitigating severe shocks like climate change, disease, cyberattacks, and disinformation campaigns. Therefore, resilience must be a priority for every presidential administration regardless of political party.

The US and its allies and partners must revivify the political warfare techniques of the Cold War and update them with today's technologies to exploit the resilience vulnerabilities inherent in authoritarian states while bolstering their own resilience to both systemic shocks and adversary IW campaigns. In an era of crises and strategic competition, victory will go to the states that are resilient. And while democracies and autocracies have different vulnerabilities, the need for greater resilience is becoming more apparent to leaders in both regime types. The US and its allies and partners must be prepared to play aggressive offense while maintaining a robust defense to win without fighting.

Notes

1. Borghard, "A Grand Strategy."
2. Foreign internal defense (FID) encompasses programs to protect "... society from subversion, lawlessness, insurgency, violent extremism, terrorism, and other threats to society." Although the US employs these programs to bolster the governments of partners and allies, some of these measures would also increase the resilience of the United States against disinformation campaigns. See, The Joint Staff, "Joint Publication 3-22: Foreign Internal Defense," IX.
3. Klare, *All Hell Breaking Loose*, 8–12.
4. Vahlsing, "Quantifying Risks."
5. Borghard, "A Grand Strategy."
6. White House, "Presidential Policy Directive."
7. White House, *National Security Strategy*, 14.
8. White House.
9. Borghard, "A Grand Strategy."
10. Sliwinski, "Resilience," 179.
11. When considering definitions of resilience in a national context, one often finds Zolli and Healy's central idea of "preserving adaptability" cited as foundational. This includes within NATO and US government documents. For examples, see Hodicky et al., "Dynamic Modeling for Resilience Measurement," Christie and Hunter, "NATO and Societal Resilience," and Van Metre's, "Fragility and Resistance."
12. Zolli and Healy, "Resilience: Why Things Bounce Back," 7.
13. Giske, "Resilience in the Age of Crises."
14. Munzo, Billsberry, and Ambrosini, "Resilience," 182–183.
15. Haenle, "China's Zero COVID Policy."
16. Cave, "How Australia Saved Thousands of Lives."
17. National Intelligence Council, "Global Trends," 65.
18. National Intelligence Council, 61.
19. National Intelligence Council.
20. Barnett, "Congress Must Recognize the Need"; and Kimhi and Eshel, "Individual and Public Resilience."
21. Giske, "Resilience in the Age of Crises."
22. Friedman, "The Pandemic Is Revealing."
23. Klare, *All Hell Breaking Loose*, 8.

Tools of Resilience

24. Klare, 12.
25. World Bank, "Climate Explainer."
26. Dew, "Fighting for Influence."
27. Pascu and Vintila, "Strengthening the Resilience."
28. Ucko and Marks, "Redefining Irregular Warfare."
29. Conley, Mina, Stefanov, and Vladimirov, *The Kremlin Playbook*.
30. Williams and McCulloch, "Truth Decay and National Security."
31. Fukuyama, "The Thing That Determines."
32. Cave, "How Australia Saved Thousands of Lives."
33. Canetti et al., "What Does National Resilience Mean," 505.
34. Johnson, Dowd, and Ridgeway, "Legitimacy as a Social Process."
35. Johnson, Dowd, and Ridgeway.
36. Bargués and Morillas, "From Democratization to Fostering Resilience."
37. Ballada et al., "Bouncing Back from COVID-19."
38. Pascu and Vintila, "Strengthening the Resilience," 55–66.
39. Canetti et al., "What Does National Resilience Mean," 507.
40. Canetti et al.
41. Petit, "Can Ukrainian Resistance Foil a Russian Victory?"
42. Gerschewski, "The Three Pillars of Stability."
43. Ngoun, "Adaptive Authoritarian Resilience."
44. Cassani, "Social Services to Claim Legitimacy."
45. Cacioppo, Reis, and Zautra, "Social Resilience," 48.
46. Bonnano, "Loss, Trauma, and Human Resilience," 25.
47. Bonnano et al., What predicts psychological resilience?"
48. "State Resilience Index Annual Report 2022."
49. Wagnild and Young, " Resilience Scale."
50. Smith et al. "The Brief Resilience Scale."
51. Maclean, Cuthill, and Ross, "Six Attributes," 149.
52. American Psychological Association, "Building Your Resilience."
53. Riess, "Institutional Resilience."
54. OECD, *Building Trust and Reinforcing Democracy*.
55. Linden, Roozenbeek, and Compton, "Inoculating Against Fake News."
56. Hodicky et al., "Dynamic Modeling," 2639.
57. Federal Emergency Management Agency, *National Response Framework* 2019, p.15.
58. Zolli and Healy, "Resilience: Why Things Bounce Back," 130.
59. Chandra et al., "Building Community Resilience"; and Fund For Peace, "State Resiliency Index Annual Report 2022."
60. Fund For Peace, "State Resilience Index Annual Report 2022."

61. Walklate, McGarry, and Mythen, "Searching for Resilience."
62. Chandra et al., "Building Community Resilience."
63. Maclean, Cuthill, and Ross, "Six Attributes," 151.
64. Kwok et al., "What Is 'Social Resilience'?"
65. Keck and Sakdalpolrak, "What Is Social Resilience?"
66. Chandra et al., "Building Community Resilience."
67. Norris et al., "Community Resilience as a Metaphor."
68. Walpole, Loerzel, and Dillard, "A Review."
69. Pfefferbaum et al., "The Communities Advancing Resilience Toolkit."
70. Pfefferbaum et al., "Assessing Community Resilience."
71. Olszewski, Liu, and Cunningham, "Survey of Federal Community," 30.
72. Olszewski, Liu, and Cunningham.
73. Walklate, McGarry, and Mythen, "Searching for Resilience."
74. Obrist, Pfeiffer, and Henley, "Multi-Layered Social Resilience."
75. Stewart, Kolluru, and Smith, "Leveraging Public-private Partnerships."
76. Stewart, Kolluru, and Smith.
77. Busch and Givens, "Achieving Resilience in Disaster Management," 19.
78. White House, "Presidential Policy Directive."
79. Kimhi et al., "National Resilience."
80. Friedland, "Introduction"; and Kimhi et al., "Community and National Resilience."
81. Canetti et al., "What Does National Resilience Mean"; and Kimhi et al., "Community and National Resilience."
82. Chasdi, "A Continuum of Nation-State Resiliency," 481.
83. NATO, "Resilience, Civil Preparedness and Article 3."
84. Borghard, "A Grand Strategy."
85. Fiala, Smith, and Löfberg, *Resistance Operating Concept (ROC)*.
86. Sitaraman, "A Grand Strategy of Resilience."
87. NATO, "Resilience, Civil Preparedness and Article 3."
88. NATO.
89. Spaulding et al., "Innovation for Resilience."
90. Ballada et al., "Bouncing Back from COVID-19."
91. National Intelligence Council, "Global Trends," 68.
92. Boras et al., "Understanding Societal Resilience."
93. Sitaraman, "A Grand Strategy of Resilience."
94. Zekulić, Godwin, and Cole, "Reinvigorating Civil–Military Relationships."
95. International Federation of Red Cross and Red Crescent Societies, "IFRC Framework for Community Resilience."

Tools of Resilience 191

96. Dowd and Cook, "Bolstering Collective Resilience in Europe."
97. Jensen and Caswell, "Deterrence by Resilience."
98. Dowd and Cook, "Bolstering Collective Resilience in Europe."
99. Dowd and Cook.
100. Jensen and Caswell, "Deterrence by Resilience."
101. Dowd and Cook, "Bolstering Collective Resilience in Europe."
102. Sandeman and Hall, "NATO's Resilience."
103. Prior, "Resilience," 74.
104. Sandeman and Hall, "NATO's Resilience."
105. Greitens, "Xi's Security Obsession."
106. Leonard, "China Is Ready."
107. Gui, "Moving toward Decoupling," 57.
108. Cha, "How to Stop Chinese Coercion."
109. Sitaraman, "A Grand Strategy of Resilience."
110. Gui, "Moving toward Decoupling."
111. Dowd and Cook, "Bolstering Collective Resilience in Europe."
112. United Nations Office for Disaster Risk Reduction, "UNDRR Homepage."
113. Sitaraman, "A Grand Strategy of Resilience."
114. Quinlan et al., "Measuring and Assessing Resilience," 685.
115. Vahlsing, "Quantifying Risks."
116. Borghard, "A Grand Strategy."
117. Washington Post Editorial Board, "We Must Learn."
118. World Health Organization, "Fighting COVID-19."
119. Hlavka and Rose, "Unprecedented by most measures."
120. Walmsley et al., "Macroeconomic Consequences."
121. Dowd and Cook, "Bolstering Collective Resilience in Europe."

CHAPTER 9

MEASURING SUCCESS

You can't improve what you can't measure.[1]

Peter F. Drucker

INTRODUCTION

This book has framed the objectives of IW as power, influence, and legitimacy in furtherance of American grand strategy and provided an overview of various categorical tools of statecraft along with recommendations for how they might be better employed. Nonetheless, the key question, "How does a country obtain and sustain legitimacy, power, and influence?" remains unanswered. To win in an era of strategic competition and crises, a state bolsters its power, influence, and legitimacy while diminishing those of its competitors. However, the profoundly political nature of IW and the relative intangibility of legitimacy and influence compared to power make it difficult to measure progress.[2] But without clear measures, it is impossible to know if one is succeeding, particularly in a competition that may endure for decades. It may be tempting to confuse effort with impact and measure that which is more easily quantified.[3]

Despite the analytical rigor policymakers have invested in developing strategies, comparatively little attention is paid to measuring progress toward their stated goals. Often, clear measures of success are conspicuously absent. In other instances, policymakers use the wrong metrics, usually because they were readily available at the time. Examples include the tally of casualties in the Vietnam War that was used as a proxy for the success of American strategy[4] and the burn rate of economic assistance in America's most recent wars that was used as a measure of counterinsurgency progress.[5] Neither correlated with a successful outcome but were used because they were readily available.

Too often, one counts what one can, which may or may not help to achieve one's objectives. To avoid this mistake, it is critical not only to determine but also to define the measurements that actually correlate with success. With that in mind, this chapter summarizes common measurement approaches, reviews the most relevant indexes and frameworks, and—finding these to be lacking—proposes a mosaic approach to measuring progress. It concludes with recommendations for amplifying the importance of measurement in orienting both domestic and foreign policy towards the same strategic objectives.

Measurement Approaches

There are multiple approaches to measuring success, each with its own merits and limitations. A common one is to articulate long-term foreign policy objectives and then measure progress relative to their achievement. Given that strategies seek to align available means with desired ends, the achievement of those ends is an intuitive metric by which to measure progress. However, this approach is problematic. First, specific long-term objectives vary across administrations and shift with changes in the international system or shocks like the bombing of Pearl Harbor or the 9/11 terrorist attacks. Second, grand strategies sometimes endure for decades, rendering measurement against ultimate outcomes unfeasible.

Measuring Success 195

For example, the Cold War containment strategy lasted for more than forty years, oscillating between symmetric and asymmetric approaches depending on the geopolitical circumstances.[6] Finally, various exogenous and unpredictable factors can affect success; for example, the Soviet invasion of Afghanistan (and international support to proxies) accelerated its demise, thus making US political warfare more successful. Indeed, "good" strategies can fail, and "bad" strategies can succeed.[7]

Rather than articulating specific intermediate or long-term objectives, we use our three dependent variables—power, influence, and legitimacy— as enduring foci around which IW campaigns are built. Simply (re)stated, power is the ability to coerce, influence is the ability to convince, and legitimacy is the collective belief that something is right. Drawing upon the US military's vocabulary,[8] measures of effectiveness (MOEs) are used to evaluate the impact and success of policies or strategies. MOEs indicate if the chosen policies (or strategies) actually facilitate the achievement of one's objectives. Therefore, to know if American IW approaches are succeeding or failing, analysts must identify MOEs for power, influence, and legitimacy. Improvements would demonstrate that the US and its allies and partners are pursuing the right approaches.

However, measuring power, influence, and legitimacy in isolation is insufficient. There is a direct link from the military, economic, information, and resilience tools of statecraft (the independent variables) to achieving relative power, influence, and legitimacy (the dependent variables). Therefore, assessing the degree to which programs and policies are implemented requires considering measures of performance (MOPs) or actions tied to task accomplishment. As such, when using MOEs to assess whether an IW strategy has increased power, influence, or legitimacy, MOPs have to be used to evaluate whether investments in military, economic, information, or resilience statecraft have occurred as directed.

Using MOEs centered on increasing power, influence, and legitimacy alongside MOPs derived from the outputs of military, economic, information, and resilience initiatives, a cohesive dashboard has been formulated. The Irregular Warfare Dashboard links targeted outputs (MOPs) across these domains—military, economic, information, and resilience—to desired outcomes (MOEs) related to power, influence, and legitimacy. To implement this approach, the following section reviews the existing research on measuring strategic success (specifically as it pertains to power, influence, and legitimacy), highlighting the value that such sources provide while also critiquing their methodologies and relevance to the concepts discussed in this book.

Survey of Scholarship

The research on measuring power, influence, and legitimacy can be divided into two types: indexes and frameworks. The most immediately useful sources are indexes, which are composite statistics that measure changes in a representative group of data points over time. Such indexes are usually updated regularly, subject to the oversight of methodological transparency and replication, and offer a clear and accessible standard for tracking progress or lack of it. These indexes may be limited by the ways their data points are defined, which are usually unique to each source and, therefore, easy to misinterpret or misapply. The second category—frameworks—sets forth helpful analytical approaches to conceptualize how one might measure particular variables to achieve a certain outcome. However, frameworks stop short of quantification. Perhaps the most comprehensive framework is RAND's "Understanding A New Framework for Competition," which includes four components to assess competition between major powers: 1) the overall context for the competition, 2) national power and competitiveness, 3) international position and influence, and 4) the shape and standing of bilateral competitions.

Both indexes and frameworks address power, influence, and legitimacy in different ways, but neither treats them as related concepts, regarding them to be mutually exclusive. Power and influence are often conflated or combined, while legitimacy is subsumed within measures of resilience or governance. Similarly, though existing indexes address some of the

Measuring Success 197

aforementioned independent variables, none address all of them, nor do they align themselves exactly to the book's framing of military, economic, information, and resilience toolsets. Therefore, this chapter mines the literature for relevant measurements upon which a dashboard for IW can be customized.

Indexes

Most indexes focus on hard power and use concrete indicators such as GDP and defense budgets, which lend themselves well to quantification. Perhaps the most famous example is the frequently cited Correlates of War (COW) project and its Composite Index of National Capabilities (CINC). The CINC uses six broad measures—military expenditure, military personnel, energy consumption, iron and steel production, urban population, and total population—to estimate a state's ability to wage conventional war.[9] These measures fail to consider qualitative elements of power (e.g., technological innovation, national will) and omit more difficult-to-measure variables such as influence and legitimacy.

More recent metrics like the Foreign Bilateral Influence Capacity (FBIC) have sought to fill the gap by including relative factors. In particular, rather than examining individual states in isolation, the FBIC uses 213 countries with units of analysis comprising directed dyad relationships (unique combinations of two states) and multiple measures across economic, security, and political dimensions to measure influence between every pair of states from 1960 to 2020.[10] Hence, the FBIC captures both the volume of total interactions (bandwidth) and the relative resource flows (dependence) between countries. For example, if policymakers wanted to know how much US influence has expanded over time globally, in Africa, or specifically in Tanzania, the FBIC could quickly show those progressions. Furthermore, the FBIC can break down those aggregate metrics into bandwidth and dependency, and measure for economic, military, and political influence. Thus, it is possible to better understand the relational aspects of both power and influence rather than just comparing power using resource-focused indexes.

FBIC measures indicate that although China has gained significant influence over many countries in the last twenty years, the US has not lost as much influence as some would believe, and many US allies have also gained influence.[11] When researchers, policymakers, or strategists use the FBIC to disaggregate the influence relationship into the various measures, they can better understand what tools of statecraft to use in order to refine strategies. The ability to isolate how particular components negatively or positively impact influence is particularly useful to policymakers, as it may provide insight as to whether particular policies resulted in their intended effect.

Other indexes recognize key differences between power and influence and attempt to measure both. For example, the Lowy Institute's Asia Power Index, albeit limited to the Indo-Pacific, provides an interesting model for quantifying power and influence separately.[12] The index contains four indicators for resources (economic capability, military capability, resilience, and future resources) which closely align with the book's conception of power. It also includes four indicators for influence: economic relationships, defense networks, diplomatic influence, and cultural influence.

However, there are issues with this approach. One issue is that the common association between power and latent resources, on the one hand, and between influence and thematic measures, on the other, does not match our recommended definitions of power and influence. Latent resources that cannot be mobilized in the near term do not equal, "the ability to coerce." While our framework will similarly divide measurements across power and influence as variables, we will not solely associate latent resources with power and the specific thematic measures with influence. The operational definitions in these indexes are not a perfect match for our conceptual definitions.

There are also a few indexes that measure legitimacy. For example, the World Bank's Worldwide Governance Indicators index combines data on governments and their citizens, aggregating factors such as voice and

Measuring Success 199

accountability, political stability, government effectiveness, regulatory quality, rule of law, and control of corruption, as components of state legitimacy.[13] Freedom House's *Freedom in the World* report tracks global political rights and civil liberties, which may be proxies for legitimacy in democracies, but it does not provide a useful measure of autocratic legitimacy, which operates along different dynamics (e.g., economic fulfillment instead of political representation).[14] Whereas ideological or regime-type related measures such as Freedom House and Polity[15] are important measures for understanding values affinity between countries, policymakers must be careful to not assume that autocratic regimes are intrinsically lacking in legitimacy, even if US values lead in that direction.

The Fund for Peace's State Resilience Index (SRI) and Fragile States Index (FSI) represent two novel approaches to integrating domestic and foreign indicators and measuring power, influence, and legitimacy. As chapter 8 discusses, the SRI measures state-level resilience in 154 countries across seven broad societal pillars: inclusion, social cohesion, environment/ecology, individual capabilities, state capacity, economy, and civic space, which in turn are broken into thirty-nine sub-pillars.[16] This index is helpful for understanding national resilience and can be disaggregated to better appreciate how certain resilience indicators impact state power, influence, and legitimacy.

The US currently performs well in the SRI, except for noticeably lower scores in environment/ecology and social cohesion. In examining the sub-pillars of social cohesion, the US score is degraded by lagging confidence in national institutions.[17] This provides actionable information, demonstrating the utility of metrics in assessing and reassessing US policy initiatives. In this case, US policymakers with an interest in winning without fighting should focus on (re)building trust in national institutions. Weak social cohesion leaves the US vulnerable to competitor IW approaches.

Finally, the FSI annually ranks 179 countries based on twelve key political, social, and economic indicators and over one hundred sub-indicators that impact their levels of fragility.[18] Though the focus on state fragility has diminished, many FSI indicators are applicable to IW objectives. For example, its economic indicators relate to state power, its social and cross-cutting indicators include proxies for influence, and its cohesion and political indicators provide measures of legitimacy. Furthermore, because the data is collected and published annually, it is a useful way for states to see how they have progressed or declined over time and how they compare to other states. Though this book focuses on strategic competition, examining the world through the lens of fragility is also useful. Significant dysfunction exists not only in poorer and weaker countries but also in richer and more powerful ones. In the US, for example, hyperpartisan domestic politics characterized by growing Group Grievance and Factionalized Elites—two key fragility indicators—along with other indicators of economic, social, and demographic disruptions, point to weaknesses where policymakers should focus their efforts.[19]

Many of the aforementioned indexes and their indicators provide a great starting point for understanding a state's strengths and weaknesses. Additionally, using these indexes to monitor progress or deterioration over time and to compare states is incredibly valuable. The greatest advantages of using such indexes are their accessibility and the breadth of their data, most of which are kept up to date. However, they are limited by their definitions of concepts like power, influence, legitimacy, governance, and resilience, which heavily shape the measurements but sometimes embody black boxes left to the reader's interpretation.

Frameworks
Another method of measurement is to use analytical frameworks that suggest ways to conceptualize progress. Generally, these sources critique existing measurements or advance their own. They are relatively more nuanced in their treatment of different indicators and provide a useful starting point. However, unlike the indexes, they do not operationalize concepts or regularly update their data, making them less useful to policymakers.

A variety of scholars advance arguments about measures of power, influence, and legitimacy. Their methodologies vary considerably and are difficult to calculate because these definitions are abstract, largely unobservable, and context-dependent.[20] Similar to some of the aforementioned indexes listed, most scholars measure power by using natural resources, GDP (or other measures of wealth), and military assets or capacity.[21]

Other scholars have advanced more nuanced arguments about metrics, which is the primary advantage of the literature on concepts. For example, Michael Beckley critiques the COW's CINC index, arguing that it is based on gross indicators that ignore costs and, therefore, mischaracterize relative power. Instead, he suggests a hybrid approach that tempers resource-based measures with historical outcome data and the unmeasured costs of certain resources (e.g., large populations).[22] Beckley's approach provides a more generalizable and historically valid measure that could prevent policymakers from overestimating the power of strategic competitors. This is important because estimations of relative power are at least partially responsible for policymakers' increased interest in expanding military power. This could be at the expense of US influence and legitimacy efforts, which are equally important in IW. Nevertheless, estimates of military power, even when carefully assessed in their situational and relative contexts, have sometimes proven unreliable when they omit critical human factors, such as will and morale, that are difficult to measure. Ukraine's military power, for example, was vastly underestimated by security experts before Russia's invasion in February 2022.[23]

Similarly nuanced, RAND's *Understanding a New Era of Strategic Competition* contextualizes metrics of success for the US, China, and Russia by weighing them against their respective objectives.[24] This framework comprises four components: overall context for the competition, national power and competitiveness, international position and influence, and bilateral competitions. The second component matches this book's definition of power more closely and can be measured through multiple variables. These include overall productive capacity, capacity to generate spending power and discretionary resources, quality of national institutions, and military resources and capabilities. Meanwhile, the third component (international position and influence)[25] matches this book's definition of influence and can be measured through economic posture and engagement, alignment of allies and partners, ideological influence, and military engagement.[26] The report also helpfully examines domestic factors, such as national competitiveness (including governance effectiveness), alongside international influence, which includes economic, military, and ideological components.

The degree of nuance in the conceptual literature also allows for greater debate on legitimacy—the belief that government leaders have the right to govern—which is extremely useful for our purposes.[27] When that belief is shared by many individuals or groups that hold power and influence, legitimacy produces collective effects.[28] As M. Stephen Weatherford explains, policymakers can use carefully constructed and implemented public opinion polls as well as measures of government responsiveness to citizen demands as proxies for legitimacy.[29] Though legitimacy is most often associated with states and their domestic audience,[30] external or international legitimacy (i.e., the degree to which states are recognized and respected by other states[31]) is also important in IW. Although obtaining accurate polling from authoritarian regimes can be difficult, accurate readings of legitimacy are still possible in most countries where the US and its allies compete. For example, the Asian Barometer Survey,[32] which began in 2001, has consistently found that the region's autocracies (like China and Cambodia) far outperform democracies (like South Korea and Japan) in institutional trust (a proxy for regime legitimacy).[33]

Critique of the Literature
Although indexes and frameworks provide helpful starting points for measuring IW success, several key flaws prevent their direct adoption and application. First, within each of the types, the concepts of power, influence, and legitimacy lack differentiation. Power and influence are

often conflated or combined, whereas legitimacy is subsumed within measures of resilience or governance.

Since this study defines these dependent variables as related but mutually exclusive concepts, any framework used in this book should share the analytical clarity of these definitions, which should be as discrete as possible. However, in the literature, many metrics overlap—for example, the same productive capacities, latent resources, or economic dependencies that allow a country to shape other states' behavior (power) often also allow that country to shape other states' perceptions (influence).

Some sources, like the Lowy Center's Asia Power Index and RAND's *Understanding a New Era of Strategic Competition*, try to address this issue by defining power as latent resources, relationships, or leverage and defining influence as the active utilization of those latent factors. However, this distinction does not accord with this study's definitions since both latent and active factors shape power, influence, and legitimacy. For example, GDP and resource endowments are crucial to the development of economic power, but so are domestic policies that increase productivity. Therefore, the literature does not provide an applicable conceptual framework or analytical index for distinguishing among the aforementioned dependent variables, which complicates developing measures of effectiveness.

Relatedly, the aforementioned independent variables overlap in the literature. For example, the same productive capacities that indicate a strong economy also contribute to military and resilient power, making it difficult to ascertain performance measures specific to each of the four domains of our strategy. Moreover, the literature often associates military and economic tools with power, information tools with influence, and resilience tools with legitimacy. However, as demonstrated throughout the book, each of the independent variables affects each of the dependent variables in different ways. Therefore, it is critical to ensure that military, economic, information, and resilience tools are mutually exclusive (to the degree possible), allowing for determining measures of performance for each of them.

The literature is divided between broad frameworks and specific indexes. Though frameworks provide analytical nuance, they fail to match the mutually exclusive definitions of independent and dependent variables. While indexes track progress over time, their quantitative indicators often diverge from the broader concepts they intend to measure. Thus, neither the frameworks nor the existing indexes match the objectives or tools for our strategic approach. Therefore, a mosaic approach is proposed, using existing strengths indexes and frameworks to connect military, economic, information, and resilience tools to power, influence, and legitimacy.

Toward an Irregular Warfare Dashboard
The proposed mosaic approach for measuring progress relies on the mutually exclusive definitions of both the independent variables (aligned as rows) and the dependent variables (aligned as columns). Drawing upon the relevant portions of the aforementioned sources, a comprehensive set of metrics has been developed to measure progress in the use of those tools.

The result is a conceptual dashboard in which measures along military, economic, information, and resilience domains indicate the impact of the use of their respective tools on relative power, influence, and legitimacy. Each suggested indicator is discrete, measurable, and tailored to the proposed strategy.

This purpose of the Irregular Warfare Dashboard is to visualize the impact of various interventions across domains—economic, military, information, and resilience—as policymakers monitor the success or failure of a particular IW strategy.

Measuring Success   205

Power

Description/Objective:
- Greater ability to coerce than competitors

Military metrics:
- Defense Spending
- Size and range of nuclear arsenal
- Size and quality of military force
- Military capabilities and capacity
- Simulations and wargame outcomes
- Magazine depth and defense industrial base

Economic metrics:
- GDP
- Labor productivity
- Total trade/relative trade
- Technology exports
- World digital competitiveness ranking
- Patent applications
- Energy production and consumption

Information metrics:
- Access to truthful information
- Access to internet & VPNs
- Effectiveness of information/intel orgs
- Percent of population information literate
- Openness
- IT Competitiveness

Resilience metrics:
- Environment and Ecology
- Demographics (gender, age, etc.)
- Geographic vulnerability (landlocked, coastal, size, neighbors, chokepoints)
- Individual capabilities

Influence

Description/Objective:
- Greater ability to convince international community, allies, partners, and fence sitters than competitors

Military metrics:
- Basing, access, and overflight
- Defense-related agreements (Logistics, Intel...)
- Multinational exercises/partner participation
- PME, IMET, FMS, and FMF
- Military exchange programs (e.g., SPP)
- KLEs & Coordination mechanisms

Economic metrics:
- Trade dependence
- Strength of currency
- Economic Assistance
- FDI
- Trade Agreements
- Global Innovation Index

Information metrics:
- Market share of int'l students studying abroad
- Information penetration abroad
- Prevalence of artistic content creation
- Index of ideological affinity
- Level of diplomatic representation
- Membership in information sharing institutions
- Commitment to free and open internet

Resilience metrics:
- State Capacity
- Social Cohesion
- Relative support provided to states in crisis
- Responsiveness of int' institutions to crisis

Legitimacy

Description/Objective:
- Domestic and global population perceives US actions and institutions are more right and appropriate than competitors

Military metrics:
- Overseas infrastructure for humanitarian operations
- Perception of military interventions
- Commitment to international military agreements
- Perceptions of military quality

Economic metrics:
- International commitment to enforcing sanctions
- Prevalence of currency use
- Perception of U.S. economic system/use of economic tools
- Quality of economic assistance

Information metrics:
- Public attitudes towards major powers
- Information perceived as more credible
- Actions taken in information domain perceived as appropriate
- Perception of competitor actions in information domain seen as inappropriate

Resilience metrics:
- Civic Space and inclusion
- Confidence in national institutions
- Perception of corruption
- Perception of human rights

Figure 9. Irregular Warfare Dashboard.

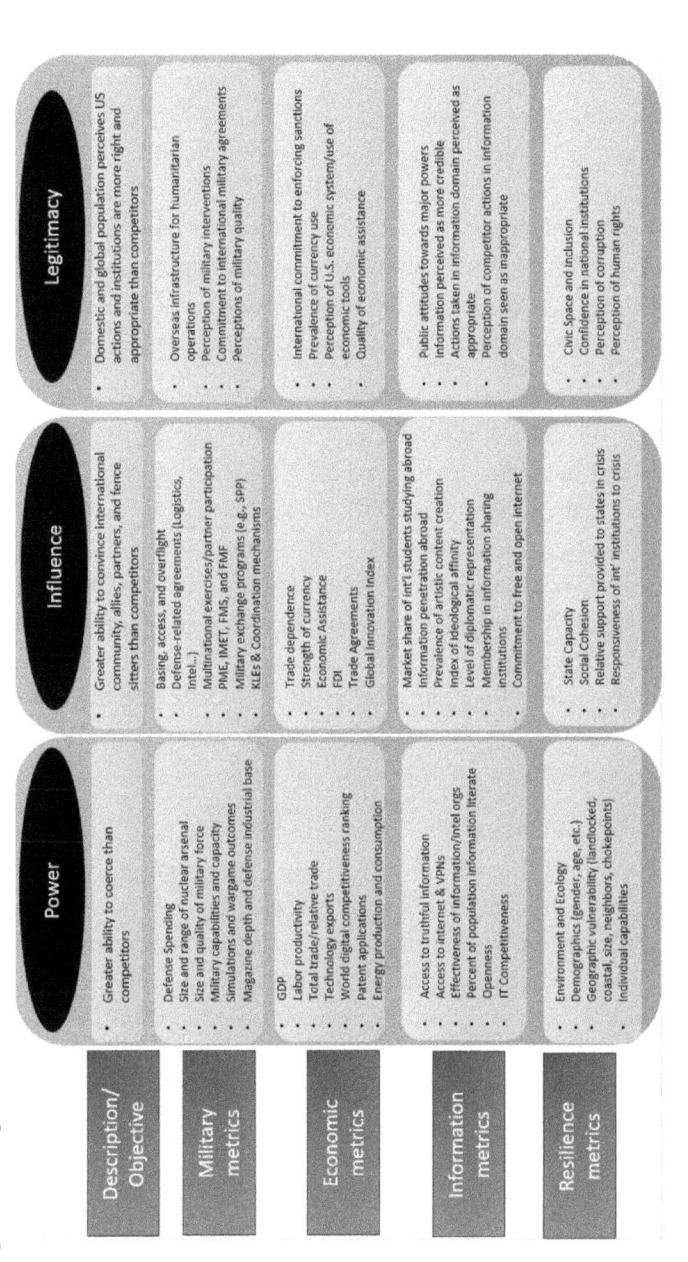

There are two key principles when using this dashboard. First, all metrics are relative. None are absolute values because they mean nothing in isolation. Thus, the recommended measures of effectiveness for power, influence, and legitimacy must always be represented in comparison to those of strategic competitors, particularly China and Russia, and any state that would support efforts to destabilize or replace the existing international order. However, relativity is not only important with respect to adversaries; it is also essential to continually assess how the US is performing longitudinally (over time) in order to determine progress.

Second, the Irregular Warfare Dashboard orients toward collective capabilities. In other words, defining and measuring success should be understood not only in terms of the US alone but also in terms of allies and partners. In an era dually defined by strategic competition and transnational crises, the US cannot or should not pursue every objective unilaterally—there are simply not enough resources. Therefore, for each of these metrics, it is essential to calculate composite scores for the US and its allies. However, by the same token, analysts should also consider grouping adversaries together—just as the US seeks to build coalitions of like-minded partners, so too will its competitors. In this regard, the US, thankfully, maintains a distinct advantage.

The mosaic approach is also meant to be flexible. Each metric can be measured and calculated in different ways depending on the purpose of the inquiry. For example, US economic assistance to partners can be measured in multiple ways: relative to adversaries, to determine the strength of the US economic toolset; as a percentage of a specific target partner's overall economic aid, to understand the extent of American economic power and influence over that state; and over the previous few years, to determine performance trendlines. There is no exclusive way to read or use the dashboard, which provides an abundance of opportunities for analysis.

There are, however, key limitations. Despite attempting to create discrete variables, many metrics still overlap. Latent geographic factors, such as economic chokepoints and natural resilience to climate change, affect military competition. However, geography is only listed under "resilience metrics and power." Economic institutions, currently at the intersection of "economic metrics" and "influence," provide leverage for pressuring another state's behavior (power) as well as its perceptions (influence). The same problem that plagues the literature also diminishes the proposed dashboard; hence, we caution against overreliance on any single tool.

There is another fundamental limitation to measurement and the blind use of cognitive models that one must bear in mind. This is best reflected in Goodhart's law, which notes that "when a measure becomes a target, it ceases to be a good measure."[34] As chapter 2 emphasizes, the US is endowed with a strategic culture characterized by a bias for action coupled with an obsession with metrics and quantification.[35] Not everything can be quantified, however. Power, influence, and legitimacy defy simplistic measurement and Likert scales. Nevertheless, the use of carefully constructed measures is helpful for understanding and facilitating informed strategic choices. This is especially true if indexes are not used alone but reinforced by rigorous qualitative frameworks and rich descriptions to boost cultural understanding, tease out nuance, and generate causal mechanisms. Those who use the Irregular Warfare Dashboard should recognize that it is a decision-support tool—not a definitive answer to a particular question.

Additionally, users of the dashboard need to bear in mind that power, influence, legitimacy, and resilience may not be the ultimate measure of effectiveness in a grand strategy. It is conceivable that a country could gain greater power, influence, legitimacy, and resilience and somehow fail to use these to achieve some ultimate goal or objective—though it is difficult to imagine such a scenario.

Recommendations

Ultimately, the mosaic approach is a tool for national assessment, meant to measure the relative progress of an IW strategy over a long-term period. The following section provides a few considerations for how national security practitioners and scholars should use this tool.

Resource Research and Measurement Rigor in Partnership with Think Tanks and Universities

The Irregular Warfare Dashboard is merely a first step. There will be many difficult decisions about how to quantify MOEs and how to determine statistically significant correlations between the independent and dependent variables. The metrics themselves may change, and the dimensions of the dashboard may be expanded if additional domains or IW objectives are added. To ensure an abundance of creative thinking on this difficult topic, the government should orient Broad Agency Announcements (BAAs), Requests for Proposals (RFPs), and other research incentives toward greater scholarship on this topic. Specifically, there are two fronts to address, corresponding to the ways in which we categorized the literature. The first area to tackle is improving the technical collection of data and calculation of metrics for indexes. Second, funded research should focus on advancing the conceptual frameworks for measuring these values.

Adopt a Publicly Accessible Holistic Government-Wide Measurement Framework

The domains of national resilience, economics, and information all connect with individuals, communities, and private-sector actors who often do not concern themselves with foreign policy. At the same time, domestic factors are critical to building a foundation for projecting power, influence, and legitimacy abroad. The era of crises illustrates the need for holistic thinking about security because transnational threats impact the homeland directly. One way to tie together the two usually disparate groups of those focused on domestic policy and those focused on foreign policy and national security is to use the same measurement framework.

The IW Dashboard is a starting point; the realized version would be fleshed out to ensure that policymakers with disparate responsibilities and constituencies are on the same page. This will be quite difficult, especially since measuring outcomes can sometimes expose shortfalls that stakeholders would rather not make public. In addition, there will likely be political incentives for any governing party to downplay negative trends and highlight positive ones—and vice versa for the opposing party. Although there is good reason to update the information regularly and report it publicly to encourage accountability, this will exacerbate political pressure to manipulate the data. Therefore, an organization with broad oversight and reach and a nonpolitical status, such as the Government Accountability Office or Congressional Research Service, should be charged with regularly measuring, calculating, and updating the dashboard while circulating it within and potentially beyond the government.

Foment A Culture of Assessment
Establishing measurement tools and providing resources are insufficient to truly alter behavior—a change in culture is needed too. Creating a government-wide culture of assessment involves cultivating an environment where data-driven decision-making is prioritized, feedback is valued, and continuous improvement is seen as a collective goal. The Biden administration has taken significant steps in this direction to fortify and expand evidence-based policymaking and agency learning agendas, but these efforts can be greatly expanded.[36]

Organizations and agencies should consider how to make assessment an integral part of routine processes, project planning, and strategic initiatives. Several steps can help establish and nurture such a culture. First, leadership commitment is essential. Executive and legislative branch leadership should champion the importance of assessment and regularly communicate its significance. Second, employees across the government need training on assessment techniques, tools, and best practices. They also require access to relevant tools for aiding in data collection, analysis, and visualization.

Once measurement tools are operational, agencies should foster open communication and transparency by sharing results with relevant stakeholders and encouraging feedback and open discussion about them. Those who contribute significantly to assessment efforts should be recognized and rewarded. Finally, the government must focus on results-driven action. Assessment is not just about collecting data; it is about using data to make informed decisions. By fostering a shared belief in the value of assessment, organizations can create a culture where data-driven decision-making is the norm, and continuous improvement is a shared objective.

Notes

1. Drucker, *Classic Drucker*.
2. Cronin, "Irregular Warfare," 2.
3. Cronin, 3.
4. Gartner and Myers, "Body Counts."
5. Querze, "Can You Spend Your Way to Victory?"
6. Lucas and Mistry, "Illusions of Coherence," 45.
7. Betts, "Is Strategy an Illusion?"
8. US Department of Defense, "United States Government Compendium," 587.
9. Correlates of War, "National Material Capabilities (v6.0)."
10. Moyer et al., "Power and Influence"; and Moyer et al., *China-US Competition*.
11. Moyer et al., "IN BRIEF."
12. Lowy Institute, "Lowy Institute Asia Power Index."
13. World Bank, "Worldwide Governance Indicators."
14. Freedom House, "Freedom in the World."
15. Center for Systemic Peace. "The Polity Project."
16. Fund for Peace, "State Resilience Index: Measuring Capacities."
17. Fund for Peace.
18. Haken et al., "Fragile States Index 2023."
19. Haken et al., 37.
20. Nye, *The Future of Power*, 3.
21. Paul Kennedy, *The Rise and Fall*; Mearsheimer, *The Tragedy*, 55–138; Nye, *The Future of Power*, 25–81; and Tellis et al., "Measuring National Power," 1–33.
22. Beckley, "The Power of Nations."
23. O'Brien, "The War That Defied Expectations."
24. Mazarr et al., *Understanding a New Era*.
25. "International position and influence" is defined as the elements of global posture, power, and influence that shape a nation's relative position. See Mazarr et al.
26. Mazarr et al.
27. Dellmuth and Schlipphak, "Legitimacy Beliefs."
28. Hurd, "Legitimacy."
29. Weatherford, "Measuring Political Legitimacy," 150.

30. Gilley, "The Meaning and Measure of State Legitimacy."
31. Coicaud, "Deconstructing international legitimacy."
32. Asian Barometer Survey, "The State of Democracy."
33. Nathan, "The Puzzle of Authoritarian Legitimacy," 158.
34. What is now known as Goodhart's law is most often generalized in a quote from anthropologist Marilyn Strathern. Strathern, "'Improving ratings'"; and Goodhart, "Problems of monetary management."
35. Johnson, "Fit for Future Conflict?"
36. White House, "FACT SHEET"; and US Department of State, "Department Of State Learning Agenda 2022 2026 (2)."

CHAPTER 10

CONCLUSION: AN AMERICAN GRAND STRATEGY FOR WINNING WITHOUT FIGHTING

> War appears to be as old as mankind, but peace is a modern invention.[1]
> Henry James Sumner Maine

INTRODUCTION

The post–Cold War era is over, and perhaps even the post-post–Cold War era has ended. Regardless of the current era's name, it is clear that conflict, even chaos, is at the door and on the horizon. A warming planet, increasing resource scarcity, and competition among strategic powers for regional or global hegemony indicate that the long period of relative global peace and prosperity experienced in the second half of the twentieth century is over. Thus, a grand strategy that addresses these realities is required.

This chapter recapitulates the most salient dynamics of the current era and its implications, emphasizing the IW approaches most likely to produce the desired outcomes for the US and its allies. It then suggests an American grand strategy that addresses the threats from both domestic and international sources. The chapter concludes with a way ahead, integrating recommendations made in earlier chapters.

THE IMPLICATIONS OF AN ERA OF COMPETITION AND CRISES

Two competing and incommensurable visions animate the geopolitical landscape. The first is advanced by China, Russia, and a raft of "lesser included autocracies," who aim to foment and capitalize on the growing disorder relying on IW to make the world safe for autocrats, kleptocrats, and plutocrats.[2] This vision is characterized by substantially circumscribed individual rights and tightly controlled (mis)information flows that benefit the most powerful at the expense of everyone else. It is also characterized by fewer legal, ethical, and political restrictions against using offensive IW campaigns to diminish adversaries and sow the seeds of political discord domestically and internationally.

The second vision is internationalist; its adherents include the US, its allies, and its partners. These actors seek to maintain the foundations of the post–World War II international order, with its emphasis on multilateralism, the rule of law, economic cooperation and integration, and poverty reduction. Although increasingly imperiled by revisionist, predominantly autocratic powers, this vision remains extant for democratic states. Likewise, the values that have underpinned it remain relevant. These competing interpretations of the global order comprise the ideological contours of the current geopolitical competition and animate the ongoing strategic competition. Yet, IW approaches can and should be used by the US and its allies and partners to counter China and Russia.

The Structure of the International System

In the past few years, there has been much debate about the structure of the international system. Although most scholars believe the American unipolar moment is over, others disagree, citing America's enduring economic dominance[3] and its unmatched military capabilities as supporting evidence for this claim.[4] Although the unipolarity adherents are definitely in the minority, there is contentious debate as to whether the emerging international system is a bipolar[5] or multipolar one.[6]

As Emma Ashford and Evan Cooper posit, the emerging international order is one of growing "unbalanced multipolarity."[7] In such a system, "power is increasingly diffusing away from the superpowers toward a variety of capable, dynamic middle powers that will help to shape the international environment in coming decades."[8] Although the US and China are currently *primus inter pares* in the international system, there are many other states with enough power to wield significant influence in global affairs and chart a course separate from that of both dominant powers, or tip the scales in favor of one or the other. The best-case scenario for the international system, then, is an unstable and negative peace, defined by an absence of armed conflict rather than a positive one based on shared norms and values.[9] This is not the vision that fueled the creation of the Bretton Woods institutions, nor is it consonant with the deeply held American belief that peace is the natural order and war the aberration.

An unstable multipolar system has far different strategic implications than a bipolar one; it also increases the relevance of IW capabilities. Although the chance of war increases, the likelihood of a major power war decreases,[10] and economic power is more diversely distributed. Civil wars, small wars, and humanitarian crises will require targeted IW capabilities more than those related to conventional military intervention. Further, a grand strategy based on forming blocs such as those constructed by the US and the Soviet Union during the Cold War is less likely to be effective in the current multipolar environment. Too many options available to middle powers do not involve stark "us or them" choices. Consider the case of Huawei in Europe as a cautionary tale,[11] or Turkey's purchase of Russian air defense systems,[12] despite being a NATO member.

Dismissing this disagreement as mere academic bloviation is tempting, but the view one subscribes to has substantial implications for strategic decision-making in the coming decades. If one believes in an emerging bipolar competition with the PRC, one would logically divert resources toward preventing China's ascendancy. America's cultural preference for a bright line between war and peace, coupled with Congressional dysfunction, unsurprisingly results in more spending on conventional military power and only ad hoc efforts to improve resilience, build influence, and increase legitimacy. Most of the administration's policy statements indicate a bipolar orientation despite diverting significant resources to support Ukraine in its war with Russia and Israel against Hamas and Iranian proxies. Indeed, the 2022 National Security Strategy asserts, "The PRC, by contrast, is the only competitor with both the intent to reshape the international order and, increasingly, the economic, diplomatic, military, and technological power to achieve that objective."[13] Secretary of Defense Lloyd Austin's December 2022 statement captures the implications:

> We're aligning our budget as never before to the China challenge. We're making the Department's largest investment ever in R&D and forging stronger capabilities. And we're modernizing, training, and equipping the US military for contingencies in the Indo-Pacific. DOD is finally making fundamental and unprecedented shifts in attention and resources toward Asia.[14]

The inherent difficulty with a bipolar approach is that it over-invests limited American resources and capabilities towards one particular threat at the expense of other priorities. Another fundamental problem is that most of these expenditures are in conventional military capabilities for great-power war when IW approaches can more effectively achieve US objectives at less cost.

Conclusion 217

For instance, the DOD plans to spend over $50 Billion to produce its new Ford Class aircraft carriers, a platform that is increasingly vulnerable to China's (and other competitors') arsenal of low-cost anti-ship missiles.[15] This single platform costs only slightly less than the Department of State's annual budget.[16] In the same 2022 speech, Secretary Austin further explained: "Deterrence means seapower. So we're investing in the new construction of nine battle-force ships and our Columbia-class ballistic-missile submarines,"[17] and "we're working to be able to mobilize and deploy American troops more quickly...We've requested billions to modernize the Marines into a highly mobile expeditionary force."[18] For the same price tag, the DOD could have invested in regional force posture and IW activities focused on increasing US influence and legitimacy in the region.

Thus, the Biden administration is wagering that the emerging strategic competition will be "Cold War 2.0," with China assuming the role formerly filled by the Soviet Union. But unlike during the Cold War, there appears to be little appetite within the US government for reinvigorating the type of whole-of-government capabilities that served the country so well during the second half of the 20th century.[19] For example, the current US strategy of "integrated deterrence," which calls for the integration of all instruments of national power, is being run through the DOD.[20] Indeed, the US has prioritized increasing military spending to contain China.

Focusing on China is logical given its increasingly aggressive behavior — asserting its sovereignty in international waters,[21] consistently saber-rattling by threatening an invasion of Taiwan,[22] and using its economic power to coerce or compel the behavior of states and multinational corporations.[23] However, it overlooks China's increasing vulnerabilities: a slowing economy,[24] faltering BRI investments,[25] and a declining and restive population.[26] These all provide opportunities to decrease Chinese influence and legitimacy through non-military means (such as by offering economic incentives to counter BRI), by fomenting domestic dissent, or both. As long as the US continues to conceive its grand strategy principally through the prism of military power—as it has consistently done since the end of the Cold War—the options for employing all other instruments will remain underdeveloped, and the results will be no more satisfying than they were during the first two decades of this century.

This myopia is particularly troubling, given the differences between the international environment today and that of 1947. Despite certain similarities, the current geopolitical landscape is far more complex; there are multiple economically ascendant and nuclear-capable middle powers with distinct interests and enough national power to pursue them. Further, the rise of an array of increasingly consequential non-state actors over the last half-century—ranging from multinational corporations to super-empowered individuals to nongovernmental organizations— further complicates the picture. Given these realities, merely increasing hard power capabilities and shoring up military alliances is inadequate to gain relative power, influence, and legitimacy.

Domestic Vulnerabilities
The threat from US competitors is not the only concern that should drive strategy. There are at least three categories of domestic vulnerabilities that must be accounted for in any grand strategy. First, there are the manmade and natural crises that all countries will face. The degree of preparedness for these crises, especially relative to that of adversaries, will impact state power, influence, and legitimacy. Second, the US and its allies and partners must carefully use their finite resources to invest in the "right" things—balancing domestic and international needs. Third, domestic populations may or may not agree with their policymakers about the threats emanating from internal or external actors. They may also fall prey to malign foreign influence, which undermines government legitimacy. All of these vulnerabilities create challenges for crafting a winning grand strategy without fighting.

China's hegemonic ambitions are undisputed. At the same time, the current investments in hard power capabilities are both disproportionate and inflexible given the emergent multipolar environment in conjunction with an age of crises. Exquisite military technology was nearly useless in the face of the COVID-19 pandemic; nor will military capabilities prove useful against domestic terrorism, or severe weather events. In fact, military assets will be equally vulnerable to them unless steps are taken to improve their resilience.[27]

The effects of emerging global challenges will be multifaceted and far-reaching, as the 2022 National Security Strategy acknowledges:

> The second [strategic challenge] is that while this competition is underway, people all over the world are struggling to cope with the effects of shared challenges that cross borders—whether it's climate change, food insecurity, communicable diseases, terrorism, energy shortages, or inflation. These are not marginal issues that are secondary to geopolitics. They are at the very core of national and international security and must be treated as such.[28]

The current imbalance in American security spending is due to the decision to invest against the most dangerous scenario—a major theater war with China. This inclination is natural given American strategic culture. However, there are real consequences to anemically funding the most likely scenarios—significant disruptions, such as cascading natural disasters, another pandemic, critical infrastructure failure (either manmade or natural) and smaller conflicts over increasingly scarce resources. Being unprepared for these likely events increases vulnerabilities, and when the US and its allies are unable to respond, they leave targets open for adversary IW campaigns.

Today extreme weather events cost the American taxpayer approximately $120 billion per year, and recent estimates indicate exponential growth in their financial impact. If nothing is done to improve the situation, losses could exceed $2 trillion annually by the end of the century.[29] Comparing defense spending to domestic resilience efforts shows the stark imbalance—$2 billion is earmarked to increase domestic climate resilience[30] whereas $276 billion (in the FY2023 request) is dedicated to integrated deterrence efforts, including $34 billion to recapitalize the nuclear triad and support nuclear command, control, and communications systems.[31] Furthermore, the DOD's FY 2023 budget request for climate change was $3 billion,[32] which is 50% more than the US will spend annually on national resilience efforts that exclude the DOD.

The experiences of the first two decades of the twenty-first century diminished the American population's appetite for overseas military entanglements,[33] with only 16% in favor of an increased US military presence abroad as of August 2023.[34] Coupled with a substantial and burgeoning isolationist sentiment, particularly among registered Republicans since 2016,[35] it appears a large percentage of the population is hoping to remain disengaged from international upheavals. But, despite the seemingly growing distaste for overseas military deployments, the majority of Americans look favorably on NATO's strengthened role in Europe (perhaps because if NATO is stronger, fewer forward-deployed US troops are required), especially in the wake of the war in Ukraine; they also support continuing strong US leadership within it.[36] Further, most Americans now perceive Russia as an enemy,[37] have an overwhelmingly negative opinion about China and its role in world affairs,[38] and prefer multilateral approaches to international crises.[39] In addition to these preferences, there is an increasing appetite for using economic tools, particularly tariffs and sanctions,[40] as preferred instruments of foreign policy.

A 2023 Gallup poll found the top five American foreign policy concerns to be terrorism (49%), immigration (45%), cyberattacks (41%), drug trafficking (41%), and climate change (39%).[41] The fact that these issues are simultaneously foreign policy and domestic issues underscores the current reality: the line between such issues is getting increasingly blurred as they become more intertwined. American grand strategy must therefore focus on domestic policy as much as it does foreign policy. The homeland and its resilience are central not only to surviving the age of crises but also to emerging from it with relative power, influence, and legitimacy intact. American geography, once a protective attribute, no longer insulates the US from what happens overseas, made strikingly not only by the COVID-19 pandemic but also through Russian interference in US elections and critical infrastructure attacks.[42]

The American electorate is once again oscillating between two of its four cultural personas: Hamiltonianism and Jacksonianism.[43] As chapter 2 points out, these worldviews ask whether the country should continue to lead internationally or retreat into isolationism, relying on its advantageous geography to remain safe from external threats.[44] Though this is a cyclical phenomenon in American history, it is also a false choice. Isolationism is no longer the viable option George Washington had advocated—the US is simply too militarily, informationally, and economically interdependent. Furthermore, the looming global challenges will not spare any locale, regardless of political rhetoric or electorate preferences.

How then should the US ensure the security and prosperity of its citizenry? What role should it play in the rules-based international order? What should the US prioritize? And what are the tradeoffs? How does the US develop a grand strategy that increases its power, influence, and legitimacy in a way that aligns with its strategic culture and adheres to the international norms and values it has worked so hard to enshrine?

A Framework for Grand Strategy

The US is currently at a critical inflection point. It must define and chart the grand strategy it wishes to employ on the shifting international stage as it navigates the coming maelstrom. To succeed, the approach must continue to build power and focus on rebuilding American influence and legitimacy while employing a strategy based on collective resilience and IW.

As noted in preceding chapters, the US does not have adequate resources to fund all emergent requirements; hard decisions will have to be made about what to protect and what to risk. To accomplish this most effectively, decision-makers should frame relevant domestic issues as essential components of national security. A holistic consideration of policy issues is necessary for accurately calculating the risks associated with the tradeoffs.

Like during the Cold War, a whole-of-society approach is required. This means leveraging all formal instruments of national power, as well as the private sector, individuals, and communities—all of which are significant US strengths. IW approaches, particularly economic tools, should take center stagean approach that the American public favors.[45] Chapter 6 suggests various options, including relying less on protectionism and more on strengthening ties with allies and partners. As discussed in chapter 4, the Marshall Plan was not only about containing the Soviet Union but also building an attractive alternative to Soviet power. In a multipolar system, states have many choices. As a result, the US and its allies must work to create situations in which they are undeniably the better option because alternatives will be available. Despite its predilection for clear binary choices, American strategic culture must shift to accommodate the range of possibilities and partners.

Rather than seeking to rerun the entire Cold War playbook, the US should build a patchwork of bilateral agreements based on discrete common interests.[46] This has two principal benefits. First, it allows the US the flexibility to enter narrow agreements with a variety of actors around areas of common interest without having to establish formal, broad-based alliances predicated on collective defense. Second, it reduces fiscal pressure on the US by returning a portion of the defense burden to the middle powers themselves.

A grand strategy should also be built upon the concept of collective resilience, "the collective capacity of allied states to withstand, fight through, and quickly recover from disruption caused by military and non-military threats caused by adversaries and global challenges."[47] Like collective defense, collective resilience is based on principles such as

Conclusion 223

mutual protection, deterrence, shared responsibilities, and collaboration. When states work in tandem to build resilience among themselves, they will achieve greater effects than they would individually. Incorporating resilience at all levels—individual, community, national, and international—will be critical to state power, influence, and legitimacy.

The *2023 Fragile States Index Annual Report* notes that in the coming years, powerful states will exploit their adversaries' fragility while reducing their own vulnerabilities by building resilience to compete, concluding that[48]

> Interlocking webs of entangling relationships are redrawing the geopolitical map. This does not make fragility any less of a threat. It simply makes it more complicated.... A new consciousness of fragility [and resilience] is emerging. Leaders at all levels of power and influence who ignore this factor in their strategic calculations will be doing so at their own peril.[49]

A grand strategy built on the foundation of resilience and IW has multiple virtues. First, it reinforces communities' and states' abilities to deal with the increasing effects of global challenges. Second, it builds capacity to defend against and deter adversaries' most prevalent IW tactics: misinformation campaigns, cyberattacks, and corporate espionage. Third, it leverages a balanced use of all instruments of power rather than overrelying on hard-power capabilities, which, in addition to being tremendously expensive, have proven indecisive in recent conflicts. Working from a foundation of IW also allows the US, its partners, and its allies to lean into their cultural norms and values, reinforcing and spreading a contemporary version of the world that is rooted in the same principles that shaped the post–World War II international system.

Thus, when creating a grand strategy, it is only appropriate to frame it in terms of American strategic culture and the democratic principles that have underpinned its government since its founding. The US and its allies must use IW, which should be a first line of defense, in a balanced way, deterring and disrupting adversaries' campaigns while offering a positive, alternative vision for the future. Positive tools, such as development assistance and collaborative security cooperation, should be used initially to strengthen partners rather than to disrupt adversaries. Likewise, tools of resilience that strengthen defenses against shocks and crises and prepare vulnerable populations to resist military aggression and occupation can provide additional layers of relative strength. Although both China and Russia have significant domestic vulnerabilities that can be targeted with offensive IW, the US and its allies should avoid stoking unnecessary tension or creating a siege mentality that could justify open conflict.

A Philosophy for Winning Without Fighting

The underpinning logic of this strategy is that the US can only win by embracing the complexities of IW and building comprehensive resilience at all levels. To do so requires a clear-eyed assessment of the risks and the tradeoffs necessary to address them. Conventional military capabilities alone—or even primarily—are as insufficient for the coming era as they proved to be for the last one. American strategic and military cultures must expand the aperture and embrace the thinking of Sun Tzu and Kautilya. Their emphasis on winning without fighting, prioritizing nonmilitary tools of statecraft (such as economic tools and alliances), and embracing myriad uses of information and disinformation provide a useful framework. This approach also demands reinvestment in the capabilities——regional expertise, cultural acumen, and language skills prominent among them—that bolstered American competitive statecraft during the Cold War. This includes reprioritizing the ability to leverage and disseminate accurate information to America's strategic advantage.

Equally important, the US must capitalize on the positive aspects of its strategic culture—optimism, entrepreneurship, practicality, and commitment to progress and democracy—while avoiding its negative aspects—mirror imaging, over relying on technological solutions, attempting to quantify the unquantifiable, and thinking only in binary terms. These aspects have stymied US efforts throughout its history and will not serve it well in an era of strategic competition and crises. Finally, the US should embrace the values that have fueled the country since its founding, working toward the development of a "positive peace" in which freedom, innovation, security, and individual dignity can thrive.

RECOMMENDATIONS AND CONCLUSIONS—A WAY AHEAD

To succeed in today's complex and perilous age, the US must deliberately and holistically chart a grand strategy that aligns with these unavoidable realities. To help achieve this, recommendations include:

- reviving an American "political warfare" capability,
- increasing and sharpening the tools of military statecraft, including more focused and bureaucratically streamlined security cooperation,
- increasing the use of economic statecraft, including economic assistance and trade to provide alternatives to China's predatory BRI efforts, while more cautiously using punitive tools like sanctions,
- focusing finite resources to increase American resilience at all levels, and
- developing collective resilience internationally.

This requires being vigilant in protecting information, hardening critical infrastructure, and increasing the ability of the American population to spot misinformation. The following highlights key themes derived from the most critical recommendations, which have been put together for policymakers.

Incorporate Resilience as a Fundamental Pillar of Grand Strategy

For the US and like-minded powers to prevail in the current competition, they must approach strategic competition as a contest to increase their relative power, influence, and legitimacy. Hard power is necessary but not sufficient for success. Influence and legitimacy are equally critical in this interconnected world. The usual levers of national power are also insufficient; national resilience—the ability to withstand and rebound from shocks, whether foreseen or unforeseen—is essential not only for preserving power, influence, and legitimacy but also for projecting them internationally. Winning requires this shift in mindset.

Develop a Whole-of-Society Approach
Once individuals and communities collectively reframe how they think about security, developing strategies to enhance and protect it becomes possible. Historical and geographic influences have convinced Americans that national security is primarily about events abroad and that they can choose whether to participate. Unfortunately, as former National Security Advisor General H. R. McMaster makes clear in his "RSVP No Fallacy" of conflict, some conflicts are unavoidable, even if the American public does not realize it.[50]

This perception stands in stark contrast to that of many American partners and allies, who have the horrors of war on their homeland seared into living memory. By reframing national security for the American public to include threats ranging from wildfires, hurricanes, droughts, floods, and tornadoes to ransomware attacks and misinformation campaigns, national security can become a critical local issue for Americans. This shift will enable the development of resilience-based strategies at all levels of government. Communities are at the center of such a strategy. Their input will enhance preparedness and broaden the responsibility for security and strategic planning, all of which are necessary to implement a whole-of-society grand strategy based on resilience and IW.

Reorganize and Resource All Instruments of National Power
As noted in the preceding chapters, the current national security organizational structure is problematic. Every executive agency, except the DOD, lacks the personnel, resources, and culture needed for strategic

planning. Because of this, it is currently impossible for the US to develop sufficiently broad and robust whole-of-government strategies necessary for success in an era of crisis and competition. At a minimum, the Departments of State, Treasury, Energy, Commerce, Homeland Security, and Health and Human Services must develop robust planning and implementation capabilities. Each of these departments will play critical roles in developing the holistic, integrated, and feasible approaches required to navigate upcoming challenges successfully.

At the outset of the Cold War, the US government developed its grand strategy from the DOS Policy Planning Office. Given the current interagency realities, it is doubtful that this approach would succeed today. Nonetheless, the US government must devise an organization and process capable of developing and operationalizing an integrated approach to grand strategy based on the twin pillars of resilience and IW.

Develop a Measurement Framework and Monitor Progress

It is not enough to develop and operationalize a grand strategy. As chapter 9 explains, it is equally vital to develop and implement a measurement framework to monitor its success. Unfortunately, due to the complexity of the undertaking and the nebulousness of certain variables, it is nearly as difficult to develop a measurement framework as it is to devise the initial strategy.

This is made even more challenging by American strategic culture because of its penchant for measurements, regardless of their relevance. Often, this is because the necessary skills to develop an effective analytic framework are missing. To address this issue, it is recommended that the US government incentivize the development of robust operational analysis capabilities across all federal departments and agencies. Additionally, it should leverage existing measures developed by the private sector and through public-private partnerships. The ability to assess whether a strategy is working is as critical to its success as the ability to create it.

Chapter 9 introduces and support a mosaic approach to measuring grand strategy. This approach integrates existing indexes and frameworks while acknowledging their limitations. It serves as a starting point rather than a complete solution to the challenge of measuring grand strategy. Additionally, incorporating big data approaches and using AI tools can help develop viable and accurate methods for measurement. However, as chapter 2 points out, technology alone is not a cure-all.

Understand the Outsized Power of the Defense in the Current Environment

The adage "The best defense is a good offense" is inverted in this case. As chapter 8 explains, current global challenges increase the intrinsic value of defense. Relative power in the international system will increasingly depend upon a state's level of resilience. Resilience is fundamentally about the ability to withstand and defend against a variety of shocks, whether from extreme weather, resource scarcity, or misinformation. States with higher resilience levels will maintain greater power in the international system than those with lower levels of resilience. Therefore, in this context, a good offense is dependent on the ability to successfully defend against a wide array of threats.

It is not only resilience that requires a good defense. Any IW strategy must also effectively defend against the corrosive IW activities of adversaries. This entails building trust in democratic institutions and fixing vulnerabilities, particularly in the information and economic space, so American adversaries cannot take advantages of gaps. This, too, requires a whole-of-society approach that begins at the local level.

Aggressively Build Teams

The US cannot navigate this era alone; engaging partners and allies globally is essential not only to gain an advantage over competitors but also to withstand natural disasters. Domestic communities are crucial to developing resilience as the international community. The US should seek opportunities to build teams at every level, from small communities to multinational alliances.

The emerging multipolar environment requires the US to adjust its mindset regarding international coalitions to offset the power of adversaries like China and Russia. Instead of seeking robust and binding military alliances, the US and its allies should develop a broader set of bilateral agreements around common interests, such as economic investments and science and technology ventures aimed at finding new energy sources. These types of arrangements will afford greater flexibility and be less costly.

Be Prepared to Fight Back when Necessary
Offensively focused measures should only be used after all other options have been exhausted, but there is an important role for them. These measures are the last resort for a reason—without the foundation of power, influence, and legitimacy, offensive IW is less effective. Punitive tools still have value, but they should be used cautiously and collectively to preserve legitimacy. However, when there is a genuine need for their use, the US and its allies should be unapologetic in employing them.

CONCLUSION

In a world of increasing complexity and multifaceted challenges, the need for a more holistic, resilience-based grand strategy has never been clearer. The threats posed by adversaries, alongside cyberattacks and natural disasters, require a unified approach both domestically and internationally. Building collaborative teams, strengthening defenses, and adopting a whole-of-society approach underscores the depth and breadth of the task ahead. Though the path is complex, the US can navigate these obstacles by harnessing its collective strength and assertively defending its core values when threatened, thereby solidifying its place in the rapidly evolving world order.

Notes

1. Maine, *International Law*.
2. Leonard, "China is Ready," 119.
3. Friedman, "Still A Unipolar World."
4. Brooks and Wohlforth, "The Myth of Multipolarity."
5. For a representative argument in favor of a bipolar future, see Kupchan, "Bipoliarity Is Back."
6. For an argument in favor of emergent multipolarity, see Ashford and Cooper, "Yes, the World"; and Walt, "America's Too Scared."
7. Ashford and Cooper, "Assumption Testing."
8. Ashford and Cooper.
9. Galtung, "An Editorial," 2.
10. Walt, "America Is Too Scared."
11. Cerulus, Roussi, and Pollet, "Europe Wants to Upgrade."
12. Reuters, "Turkey's Russian Air Defense System."
13. White House, *National Security Strategy*, 8.
14. Austin, "The Decisive Decade."
15. Congressional Research Service, "China Naval Modernization."
16. United States Department of State, "FY 2023 International Affairs Budget."
17. Austin, "The Decisive Decade."
18. Austin.
19. Cleveland et al., "An American Way of Political Warfare," 2–4.
20. White House, *National Security Strategy*.
21. International Crisis Group, "Competing Visions."
22. Associated Press, "China's Foreign Minister Steps Up Threats."
23. Cha, "How to Stop Chinese Coercion."
24. Kennedy et al., "Experts React."
25. Clark, "The Rise and Fall of the BRI."
26. O'Hanlon, "China's Shrinking Population."
27. Klare, *All Hell Breaking Loose*, 37.
28. White House, *National Security Strategy*, 6.
29. Vahlsing, "Quantifying Risks."
30. FEMA, "Biden-Harris Administration."
31. Office of the Under Secretary of Defense (Comptroller)/Chief Financial Officer, *Pacific Deterrence Initiative*, 2-1.

Conclusion

32. Office of the Under Secretary of Defense (Comptroller)/Chief Financial Officer, 4–17.
33. McMann and Reid, "US Foreign Policy Tracker."
34. Kamarick and Muchnik, "One Year into the Ukraine War."
35. McMann and Reid, "US Foreign Policy Tracker."
36. Poushter et al., "Americans Hold Positive Feelings."
37. Poushter et al.
38. Gallup Polls, "China."
39. Kamarick and Muchnik, "One Year into the Ukraine War."
40. McMann and Reid, "US Foreign Policy Tracker."
41. Kamarick and Muchnik, "One Year into the Ukraine War."
42. Kavanagh and Rich, *Truth Decay.*
43. Mead, *Special Providence.*
44. Nye, "Is America Reverting to Isolationism?".
45. Kamarick and Muchnik, "One Year into the Ukraine War."
46. Ashford and Cooper, "Assumption Testing."
47. Dowd and Cook, "Bolstering Collective Resilience in Europe."
48. Haken et al., "Fragile States Index 2023," 38.
49. Haken et al., 39.
50. Amble, "How Lt Gen H.R. McMaster."

Bibliography

Abramowitz, Alan and Jennifer McCoy. "United States: Racial Resentment, Negative Partisanship, and Polarization in Trump's America." *The ANNALS of the American Academy of Political and Social Science* 681, no. 1 (January 1, 2019): 137–156. https://doi.org/10.1177/0002716218811309.

Adamsky, Dmitry. "From Moscow with Coercion: Russian Deterrence Theory and Strategic Culture." *Journal of Strategic Studies* 41, no. 1–2, (February 23, 2018): 33–60.

———. "Russian Campaign in Syria – Change and Continuity in Strategic Culture." *Journal of Strategic Studies* 43, no. 1, (January 2, 2020): 104–125.

Aggarwal, Vinod K., and Andrew W. Reddie. "Economic Statecraft in the 21st Century: Implications for the Future of the Global Trade Regime." *World Trade Review* 20, no. 2 (May 2021): 137–151. https://doi.org/10.1017/S147474562000049X.

Agnew, J., and R. Rosenzweig, eds. *A Companion to Post-1945 America.* Chichester, West Sussex: John Wiley & Sons, Ltd., 2006.

Akhtar, Shayerah, Cathleen D. Cimino-Isaacs, and Karen M. Sutter. *U.S. Trade Policy.* Washington, DC: Congressional Research Service, updated January 30, 2023.

Al-Rodhan, Nayef. "Strategic Culture and Pragmatic National Interest." *Global Policy Journal*, July 22, 2015. https://www.globalpolicyjournal.com/blog/22/07/2015/strategic-culture-and-pragmatic-national-interest.

———. "U.S. Space Policy and Strategic Culture." JIA SIPA, April 16, 2018. https://jia.sipa.columbia.edu/online-articles/us-space-policy-and-strategic-culture.

Alexseev, Mikhail. "Russia's Ethnic Minorities: Putin's Loyal Neo-Imperial 'Fifth Column.'" *PONARIS Eurasia*, August 6, 2014. https://

www.ponarseurasia.org/russia-s-ethnic-minorities-putin-s-loyal-neo-imperial-fifth-column/.

Allen, Kenneth W., John Chen, and Phillip C. Saunders. *Chinese Military Diplomacy, 2003–2016: Trends and Implications.* Washington, DC: National Defense University, 2017.

Allen, Gregory C., Emily Benson, and William Alan Reinsch. "Improved Export Controls Enforcement Technology Needed for U.S. National Security." *CSIS,* November 30, 2022. https://www.csis.org/analysis/improved-export-controls-enforcement-technology-needed-us-national-security.

Allison, Graham. *Destined for War: Can America and China Escape Thucydides's Trap?* Boston, MA: First Mariner Books Edition, 2018.

Allison, Graham, Nathalie Kiersznowski, and Charlotte Fitzek. "The Great Economic Rivalry: China vs the US." Cambridge, MA: Harvard University, Belfer Center for Science and International Affairs, March 23, 2022. https://www.belfercenter.org/publication/great-economic-rivalry-china-vs-us.

Almond, Gabriel A., and Sidney Verba. *The Civic Culture Revisited.* Thousand Oaks, CA: SAGE Publications, Inc., 1989. http://ci.nii.ac.jp/ncid/BB02421532.

Amble, John. "How Lt Gen H.R. McMaster, The New National Security Advisor, Thinks about War." *Modern War Institute.* Feb 22, 2017.

American Psychological Association. "Building Your Resilience." February 1, 2020. https://www.apa.org/topics/resilience/building-your-resilience.

Arasasingham, Aidan, and Gerard DiPippo. "Evaluating CFIUS in 2021." *Center for Strategic and International Studies,* August 9, 2022. https://www.csis.org/analysis/evaluating-cfius-2021.

Article 19. "UN: Human Rights Council adopts resolution on human rights on the internet." July 15, 2021. https://www.article19.org/resources/un-human-rights-council-adopts-resolution-on-human-rights-on-the-internet/.

Bibliography

Arquilla, John. *Insurgents, Raiders, and Bandits: How Masters of Irregular Warfare Have Shaped Our World.* Chicago, IL: Ivan R. Dee, 2011.

Arquilla, John, and David Ronfeldt. *The Emergence of Noopolitik: Toward An American Information Strategy.* Santa Monica, CA: RAND Corporation, 1999. https://doi.org/10.7249/MR1033.

Ashford, Emma, and Evan Cooper. "Assumption Testing: Multipolarity Is More Dangerous than Bipolarity for the United States." Stimson Center, October 2, 2023. https://www.stimson.org/2023/assumption-testing-multipolarity-is-more-dangerous-than-bipolarity-for-the-united-states/.

———. "Yes, the World Is Multipolar: and that isn't bad news for the United States." *Foreign Policy,* October 5, 2023. https://foreignpolicy.com/2023/10/05/usa-china-multipolar-bipolar-unipolar/.

Asian Barometer Survey. "The State of Democracy in the Post-Pandemic Era." 2023 Asian Barometer Survey Conference. https://www.asianbarometer.org/.

Asian Infrastructure Investment Bank. *AIIB Report 2021.* https://www.aiib.org/en/news-events/annual-report/2021/_common/pdf/2021_AIIBAnnualReport_web-reduced.pdf.

———. *Moonshots for the Emerging World: Building State Capacity and Mobilizing the Private Sector Toward Net Zero.* Asian Infrastructure Finance 2022, "Introduction." https://www.aiib.org/en/news-events/asian-infrastructure-finance/2022/introduction/index.html.

Asmolov, Gregory. "The Disconnective Power of Disinformation Campaigns." *Journal of International Affairs* 71, no. 1.5 (2018): 69–76. https://www.jstor.org/stable/26508120.

Asmus, Gerda, Andreas Fuchs, and Angelika Müller. "Russia's Foreign Aid Re-Emerges." *AidData's Blog,* April 9, 2018. https://www.aiddata.org/blog/russias-foreign-aid-re-emerges.

Associated Press. "China's Foreign Minister Steps Up Threats Against Taiwan." *NBC News,* sec. World News, April 21, 2023. https://www.nbcnews.com/news/world/china-foreign-minister-threats-taiwan-rcna80790.

Atlantic Council. "Adding Color to the Gray Zone: Establishing a Strategic Framework for Hybrid Conflict." https://www.atlanticcouncil.org/programs/scowcroft-center-for-strategy-and-security/forward-defense/adding-color-to-the-gray-zone-establishing-a-strategic-framework-for-hybrid-conflict/.

———. "The Longer Telegram: Toward a New American China Strategy." https://www.atlanticcouncil.org/content-series/atlantic-council-strategy-paper-series/the-longer-telegram/#national-measures.

———. "Scoping the Gray Zone: Defining Terms and Policy Priorities for Engaging Competitors Below the Threshold of Conflict." December 22, 2022. https://www.atlanticcouncil.org/content-series/strategic-insights-memos/scoping-the-gray-zone-defining-terms-and-policy-priorities-for-engaging-competitors-below-the-threshold-of-conflict/.

———. "Today's Wars Are Fought in the 'Gray Zone.' Here's Everything You Need to Know About It." February 23, 2022. https://www.atlanticcouncil.org/blogs/new-atlanticist/todays-wars-are-fought-in-the-gray-zone-heres-everything-you-need-to-know-about-it/.

Austin III, Lloyd. "The Decisive Decade: Remarks by Secretary of Defense, Lloyd J. Austin III, at the Reagan National Defense Forum." U.S. Department of Defense, December 3, 2022. https://www.defense.gov/News/Speeches/Speech/Article/3235391/the-decisive-decade-remarks-by-secretary-of-defense-lloyd-j-austin-iii-at-the-r/.

Aydintaşbaş, Asli, Pavel K. Baev, Jessica Brandt, Federica Saini Fasanotti, Vanda Felbab-Brown, James Goldgeier, Ryan Hass, Steven Heydemann, Suzanne Maloney, Michael E. O'Hanlon, Elizabeth Saunders, Constanze Stelzenmüller, and Caitlin Talmadge. "What is the fallout of Russia's Wagner rebellion?" Brookings, June 27, 2023. https://www.brookings.edu/articles/what-is-the-fallout-of-russias-wagner-rebellion/.

Babbage, Ross. *Winning Without Fighting*. Washington, DC: CSBA, July 24, 2019.

Bacchus, James. "Biden and Trade at Year One: The Reign of Polite Protectionism." *Cato Institute*, April 26, 2022. https://www.cato.org/policy-analysis/biden-trade-year-one#polite-protectionism.

Bibliography 237

Bachand, Rémi. "International economic institutions after neoliberalism: the Indo-Pacific Economic Framework for Prosperity as a blueprint?" *London Review of International Law*, lrad017, November 1, 2023. https://doi-org.proxy.library.georgetown.edu/10.1093/lril/lrad017.

Baer-Bader, Juulia. "EU Response to Disinformation from Russia on COVID-19: Three Lessons." *DGAP Commentary* 18, (June 2020). https://nbn-resolving.org/urn:nbn:de:0168-ssoar-69216-1.

Baldwin, David A. *Economic Statecraft*. Princeton, NJ: Princeton University Press, 2020.

———. "Power Analysis and World Politics: New Trends versus Old Tendencies." *World Politics* 31, no. 2, (January 1979): 161–194.

———. *Power and International Relations: A Conceptual Approach*. Princeton, NJ: Princeton University Press, 2016.

———. "Power and Social Exchange." *The American Political Science Review* 72, (1978): 1229–1242.

———. "The Power of Positive Sanctions." *World Politics* 24, no. 1, (October 1971): 19–38. https://www.cambridge.org/core/product/identifier/S0043887100001507/type/journal_article.

Ballada, Christine Joy A., John Jamir Benzon R. Aruta, Carmelo M. Callueng, Benedict G. Antazo, Shaul Kimhi, Maurício Reinert, Yohanan Eshel, Hadas Marciano, Bruria Adini, Juliano Domingues da Silva, and Fabiane Cortex Verdu. "Bouncing Back from COVID-19: Individual and Ecological Factors Influence National Resilience in Adults from Israel, the Philippines, and Brazil." *Journal of Community & Applied Social Psychology* 32, no. 3 (2022): 452–475. https://doi.org/10.1002/casp.2569.

Bamford, Tyler. "The Points Were All That Mattered: The US Army's Demobilization After World War II." The National WWII Museum, August 27, 2020. https://www.nationalww2museum.org/war/articles/points-system-us-armys-demobilization#:~:text=The%20US%20Army%20finally%20ended,684%2C000%20on%20July%201%2C%201947.

Bandeira, Luiza, Nika Aleksejeva, Tessa Knight, Jean le Roux, Graham Brookie, Iain Robertson, Andy Carvin, and Zarine Kharazian.

"Weaponized: How Rumors of COVID-19's Origins Led to a Narrative Arms Race." Washington, DC: Atlantic Council, February 14, 2021.

Bargués, Pol, and Pol Morillas. "From Democratization to Fostering Resilience: EU Intervention and the Challenges of Building Institutions, Social Trust, and Legitimacy in Bosnia and Herzegovina." *Democratization* 28, no. 7 (October 3, 2021): 1319–1337. https://doi.org/10.1080/13510347.2021.1900120.

Barnett, M.D. "Congress Must Recognize the Need for Psychological Resilience in an Age of Terrorism." *Families, Systems, & Health* 22, vol. 1, (2004): 64–66. https://doi.org/10.1037/1091-7527.22.1.64.

Bartles, Charles K. "Getting Gerasimov Right." *Military Review*, (January-February 2016): 30–38. https://www.armyupress.army.mil/Journals/Military-Review/English-Edition-Archives/January-February-2016/.

Beccaro, Andrea. "Modern Irregular Warfare: The ISIS Case Study." *Small Wars & Insurgencies* 29, no. 2 (2018): 207–228.

Becker, Abraham S. *U.S. Soviet Trade in the 1980s*. Santa Monica: RAND Corporation, November 1987.

Beckley, Michael. "The Power of Nations: Measuring What Matters." *International Security* 43, no. 2 (2018): 7–44. https://doi.org/10.1162/isec_a_00328.

Behnke, Camilla. "Putin searches for more friends at Africa summit but low turnout dampens bid for influence." NBC News, July 29, 2023. https://www.nbcnews.com/news/world/putin-searches-friends-africa-summit-low-turnout-dampens-bid-influence-rcna96599.

Bennett, W. Lance, and Steven Livingston. "The Disinformation Order: Disruptive communication and the decline of democratic institutions." *European Journal of Communication* 33, issue 2, (April 2018): 122–139.

Bensahel, Nora. "Darker Shades of Gray: Why Gray Zone Conflicts Will Become More Frequent and Complex." Foreign Policy Research Institute, February 13, 2017. https://www.fpri.org/article/2017/02/darker-shades-gray-gray-zone-conflicts-will-become-frequent-complex/.

Berger, Sam, and Alex Tausanovitch. "Lessons From Watergate: Preparing for Post-Trump Reforms." Washington, DC: Center for American Progress, July 30, 2018. https://www.americanprogress.org/article/lessons-from-watergate/.

Berger, Thomas U. *Cultures of Anti Militarism: National Security in Germany and Japan.* Baltimore: Johns Hopkins University Press, 1998.

———. "From Sword to Chrysathemum: Japan's Culture of Antimilitarism." *International Security* 17, no. 4 (1993): 119–150.

Bergmann, Max, Maria Snegovaya, Tina Dolbaia, and Nick Fenton. "Seller's Remorse: The Challenges Facing Russia's Arms Exports." Washington, DC: CSIS, September 18, 2023. https://www.csis.org/analysis/sellers-remorse-challenges-facing-russias-arms-exports.

Berman, Eli, and David A. Lake. *Proxy Wars: Suppressing Violence through Local Agents.* Ithaca, NY: Cornell University Press, 2019.

Berman, Noah and Anshu Siripurapu. "One Year of War in Ukraine: Are Sanctions Against Russia Making a Difference?" Council on Foreign Relations, February 21, 2023. https://www.cfr.org/in-brief/one-year-war-ukraine-are-sanctions-against-russia-making-difference.

Betts, Richard. "Is Strategy an Illusion?" *International Security* 25, no. 2 (Fall 2000): 5–50.

Beauchamp-Mustafaga, Nathan. "China's Military Aid Is Probably Less Than You Think." *The RAND Blog* (blog), July 26, 2022. https://www.rand.org/blog/2022/07/chinas-military-aid-is-probably-less-than-you-think.html.

Bhaskaran Pillai, Mohanan. "Indian Strategic Culture: The Debates in Perspective." *SSRN Scholarly Paper*, March 16, 2023. https://papers.ssrn.com/sol3/papers.cfm?abstract_id=3555343.

Biddle, Stephen. *Nonstate Warfare: The Military Methods of Guerillas, Warlords, and Militias.* Princeton, NJ: Princeton University Press, 2021.

Bilms, Kevin. "What's in a Name? Reimagining Irregular Warfare Activities for Competition," *War on the Rocks*, January 15, 2021.

https://warontherocks.com/2021/01/whats-in-a-name-reimagining-irregular-warfare-activities-for-competition/.

Blackwill, Robert D., and Jennifer Harris. *War by Other Means: Geoeconomics and Statecraft*. Cambridge, MA: Harvard University Press, 2016.

Blanchard, Jean-Marc F., and Norrin M. Ripsman. *Economic Statecraft and Foreign Policy: Sanctions, Incentives, and Target State Calculations*. London: Routledge, 2013.

Blitt, Robert C. "Russia's "Orthodox" Foreign Policy: The Growing Influence of The Russian Orthodox Church in Shaping Russia's Policies Abroad." *University of Pennsylvania Journal of International Law* 33, no. 2 (2011): 363–460. https://www.law.upenn.edu/live/files/142-blittboyd33upajintll3632011pdf;

Bloomfield, Alan. "Time to Move On: Reconceptualizing the Strategic Culture Debate." *Contemporary Security Policy* 33, no. 3, (December 1, 2012): 437–461.

Bodine-Baron, Elizabeth, Todd C. Helmus, Andrew Radin, and Elina Treyger. *Countering Russian Social Media Influence*. Santa Monica, CA: RAND, 2018. https://doi.org/10.7249/rr2740.

Boesche, Roger. "Kautilya's Arthasastra on War and Diplomacy in Ancient India." *Journal of Military History* 67, no. 1 (2003): 9–37.

Boone, Merlin. "Strategic Influence Operations: A Call to Action." *Journal of Public and International Affairs* (blog), August 26, 2021. https://jpia.princeton.edu/news/strategic-influence-operations-call-action.

Boot, Max, and Michael Doran. "Political Warfare," Council on Foreign Relations, 2013, https://www.cfr.org/report/political-warfare.

Bonnano, George A. "Loss, Trauma, and Human Resilience: Have we Underestimated the Human Capacity to Thrive after Extremely Aversive Events?" *Am Psychology* 59, no. 1, (January 2004): 20–28.

———, Sandro Galea, Angela Bucciarelli, and David Vlahov. "What predicts psychological resilience after disaster? The role of demographics, resources, and life stress." *Journal Consulting and*

Clinical Psychology, 75, no. 5, (October 2007): 671–682. doi: 10.1037/0022-006X.75.5.671.

Boras, Moran, Kobi Peleg, Nathan Stolero, and Bruria Adini. "Understanding Societal Resilience—Cross-Sectional Study in Eight Countries.," *Frontiers in Public Health* 10 (2022). https://www.frontiersin.org/articles/10.3389/fpubh.2022.883281.

Borghard, Erica D. "A Grand Strategy Based on Resilience." *War on the Rocks* (blog), January 4, 2021. https://warontherocks.com/2021/01/a-grand-strategy-based-on-resilience/.

Borrie, John. "Human Rationality and Nuclear Deterrence." *Perspectives on Nuclear Deterrence in the 21st Century.* https://www.chathamhouse.org/2020/04/perspectives-nuclear-deterrence-21st-century-0/human-rationality-and-nuclear-deterrence.

Bown, Chad P. "Four Years into the Trade War, Are the US and China Decoupling?" Peterson Institute for International Economics, October 20, 2022. https://www.piie.com/blogs/realtime-economics/four-years-trade-war-are-us-and-china-decoupling.

———. "US-China Trade War Tariffs: An Up-to-Date Chart." Peterson Institute for International Economics, February 14, 2020. https://www.piie.com/research/piie-charts/us-china-trade-war-tariffs-date-chart.

Boyle, Michael J., and Anthony F. Lang. "Remaking the World in America's Image: Remaking the World in America's Image: Surprise, Strategic Culture, and the American Ways of Intervention." *Foreign Policy Analysis* 17, no. 2, (2021): 1–39.

BP. "Statistical Review of World Energy 2022." https://www.bp.com/content/dam/bp/business-sites/en/global/corporate/pdfs/energy-economics/statistical-review/bp-stats-review-2022-full-report.pdf.

Bradshaw, Samantha, and Philip N. Howard. "The Global Organization of Social Media Disinformation Campaigns." *Journal of International Affairs* 71, no. 1.5 (2018): 23–32.

Brands, Hal. "The Dark Art of Political Warfare: A Primer." Washington, DC: American Enterprise Institute, February 2020. https://www.aei.org/research-products/report/the-dark-art-of-political-warfare-a-primer/.

———. *The Twilight Struggle.* New Haven, CT: Yale University Press, 2022.

———, and Jeremy Suri, eds. *The New Makers of Modern Strategy: From the Ancient World to the Digital Age.* Princeton: Princeton University Press, 2023.

———, and Jeremy Suri, eds. *The Power of the Past,* (Washington, D.C.: Brookings Institute Press, 2016).

———, and Toshi Yoshihara. "Waging Political Warfare." *The National Interest* 159 (January/February 2019): 16–26.

Brandt, Jessica. "An Information Strategy for the United States." Brookings, May 5, 2023. https://www.brookings.edu/articles/an-information-strategy-for-the-united-states/.

———. "An Information Strategy for the United States." Testimony to the Senate Foreign Relations Subcommittee on State Department and USAID Management. Washington, DC: Brookings, May 3, 2023).

———. "How Autocrats Manipulate Online Information: Putin's and Xi's Playbooks." *The Washington Quarterly* 44, no. 3 (July 3, 2021): 127–154. https://doi.org/10.1080/0163660X.2021.1970902.

———, Zack Cooper, Bradley Hanlon, and Laura Rosenberger. "Linking Values and Strategy: How Democracies Can Offset Autocratic Advances." German Marshall Fund, Alliance for Security Democracy, October 2020.

Brice, Benjamin. "A Very Proud Nation: Nationalism in American Foreign Policy." *The SAIS Review of International Affairs* 35, no. 2 (Summer-Fall 2015): 57–68.

Britzky, Haley. "Top US General Says Increased Partnership between Iran, Russia, and China Will Make Them 'problematic' for 'Years to Come.'" *CNN*, March 31, 2023. https://www.cnn.com/2023/03/29/politics/china-russia-iran-us-mark-milley-nuclear/index.html.

Brooks, Rosa. *How Everything Became War and the Military Became Everything.* New York, NY: Simon & Schuster, 2017.

Brooks, Stephen G., and William C. Wohlforth. "The Myth of Multipolarity." *Foreign Affairs,* May/June 2023. https://www.foreignaffairs.com/united-states/china-multipolarity-myth.

Bibliography

Bryant, Susan, and Mark Troutman, eds. *Resourcing the National Security Enterprise.* Amherst, NY: Cambria Press, 2022.

Bunce, Valerie J., and Sharon L. Wolchik. *Defeating Authoritarian Leaders in Postcommunist Countries.* Cambridge: Cambridge University Press, 2012.

Busch, Nathan E., and Austen D. Givens. "Achieving Resilience in Disaster Management." *Journal of Strategic Security* 6, no. 2 (2013): 1–19. http://www.jstor.org/stable/26466757.

Bush, George W. "State of the Union Address 2003." In "Selected Speeches of President George W. Bush: 2001–2008."

Byrd, William. "Why Have the Wars in Afghanistan and Ukraine Played Out So Differently?" *United States Institute of Peace* Analysis and Commentary (blog), June 23, 2022. https://www.usip.org/publications/2022/06/why-have-wars-afghanistan-and-ukraine-played-out-so-differently.

Cacioppo, John, Harry Reis, and Alex Zautra. "Social Resilience: The Value of Social Fitness with an Application to the Military." *The American Psychologist* 66, no. 1 (2011): 43–51.

Calamur, Krishnadev. "Panama's Decision to Cut Ties with Taiwan." *The Atlantic,* June 13, 2017. https://www.theatlantic.com/news/archive/2017/06/panama-taiwan-china/530106/.

Campbell, Kurt M., and Jake Sullivan. "Competition with Catastrophe: How America Can Both Challenge and Coexist with China." *Foreign Affairs,* August 1, 2019. https://www.foreignaffairs.com/articles/china/competition-with-china-without-catastrophe.

Canetti, Daphna, Israel Waismel-Manor, Naor Cohen, and Carmit Rapaport. "What Does National Resilience Mean in a Democracy? Evidence from the United States and Israel." *Armed Forces & Society* 40, no. 3 (2014): 504–520. https://www.jstor.org/stable/48609337.

Carr, Edward Hallet. *The Twenty Years' Crisis, 1919–1939: An Introduction to the Study of International Relations.* New York, NY: Harper, 1964.

Cartwright, Dorwin. "Influence, Leadership, Control." In *Handbook of Organizations*, edited by James G. March, 1-47. London: Routledge, 2013.

Cartwright, Dorwin, ed. *Studies in Social Power*. Ann Arbor: University of Michigan Press, 1960.

Cassani, Andrea. "Social Services to Claim Legitimacy: Comparing Autocracies' Performance." *Contemporary Politics* 23, no. 3 (July 3, 2017): 348–368. https://doi.org/10.1080/13569775.2017.1304321.

Cassidy, Robert M. *Russia in Afghanistan and Chechnya: Military Strategic Culture and the Paradoxes of Asymmetric Conflict*. Carlisle, PA: Strategic Studies Institute, U.S. Army War College, 2003.

Cassinelli, C.W. *Free Activities and Interpersonal Relations*. The Hague, Martinus Nijhoff, 1966.

Cave, Damien. "How Australia Saved Thousands of Lives While Covid Killed a Millions Americans." *The New York Times*, May 15, 2022. https://www.nytimes.com/2022/05/15/world/australia/covid-deaths.html#:~:text=In%20global%20surveys%2C%20Australians%20were,wearing%20masks%20and%20getting%20vaccinated.

Center for Systemic Peace. "The Polity Project." https://www.systemicpeace.org/polityproject.html.

Cerulus, Aurens, Antoaneta Roussi, and Mathieu Pollet. "Europe Wants to Upgrade its Huawei Plan." *POLITICO*, June 14, 2023. https://www.politico.eu/article/europe-plots-huawei-plan-upgrade/.

Cha, Victor. "How to Stop Chinese Coercion." *Foreign Affairs*, January/February 2023. https://www.foreignaffairs.com/world/how-stop-china-coercion-collective-resilience-victor-cha.

Chai Shaojin. "China's Nascent Soft Power Projection in the Middle East and North Africa." In *Routledge Handbook on China–Middle East Relations*, edited by Jonathan Fulton, 264-280. London: Routledge, 2021.

Chairman of the Joint Chiefs of Staff. *2022 National Military Strategy*.

Chandra, Anita, Joie D. Acosta, Stefanie Howard, Lori Uscher-Pines, Malcolm V. Williams, Douglas Yeung, Jeffrey Garnett, and Lisa S. Meredith. "Building Community Resilience to Disasters: A Way

Forward to Enhance National Health Security." Washington, DC: RAND Corporation, February 22, 2011. https://www.rand.org/pubs/technical_reports/TR915.html.

Chandra, Vikash. "How India Sees the World: Kautilya and the 21st century." *Commonwealth & Comparative Politics* 56, no. 3, (July 3, 2018): 401–403.

Charap, Samuel, Dara Massicot, Miranda Priebe, Alyssa Demus, Clint Reach, Mark Stalczynski, Eugeniu Han, and Lynn E. Davis. *Russian Grand Strategy: Rhetoric and Reality.* Santa Monica, CA: RAND Corporation, 2021. https://www.rand.org/pubs/research_reports/RR4238.html.

Chasdi, Richard J. "A Continuum of Nation-State Resiliency to Watershed Terrorist Events." *Armed Forces & Society* 40, no. 3 (2014): 476–503. https://www.jstor.org/stable/48609336.

Chase, Kerry A. "The United States and Multilateral Trade Liberalization, 1922–67." In *Trading Blocs: States, Firms, and Regions in the World Economy.* Ann Arbor: University of Michigan Press, 2005. https://www.press.umich.edu/pdf/047209906X-Ch4.pdf.

Chatzky, Andrew, and Anshu Siripurapu. "The Truth About Tariffs." Council on Foreign Relations, October 8, 2021. https://www.cfr.org/backgrounder/truth-about-tariffs.

Chau, Donovan C., ed. *Strategy Matters: Essays in Honor of Colin S. Gray.* Maxwell, AL: Air University Press, 2022.

Chen, Thomas M., Lee Jarvis, and Stuart Macdonald. *Cyberterrorism Understanding, Assessment, and Response.* New York, NY: Springer Science+Business Media, 2014.

Chidester, Jeffrey L., and Paul Kengor, eds. *Reagan's Legacy in a World Transformed*, Cambridge, MA: Harvard University Press, 2015.

"China punishes Australia for promoting an inquiry into covid-19." *The Economist*, May 21, 2020. https://www-economist-com.proxy.library.georgetown.edu/asia/2020/05/21/china-punishes-australia-for-promoting-an-inquiry-into-covid-19.

"China Spends More on Controlling Its 1.4bn People than on Defense." *Nikkei Asia*, August 29, 2022. https://asia.nikkei.com/static/

vdata/infographics/china-spends-more-on-controlling-its-1-dot-4bn-people-than-on-defense/.

Christie, Edward Hunter, and Kristine Berzina. "NATO and Societal Resilience." German Marshall Fund, July 20, 2022.

Ciddi, Sinan. "Is Turkey About to Ditch Its Russian S-400 Missile System?" *Washington Examiner*, March 29, 2023. https://www.washingtonexaminer.com/opinion/is-turkey-about-to-ditch-its-russian-s-400-missile-system.

Clark, Nadia. "The Rise and Fall of the BRI." Council on Foreign Relations, April 6, 2023, https://www.cfr.org/blog/rise-and-fall-bri.

Claude, Inis L., Jr. *Power and International Relations.* New York, NY: Random House, 1962.

Clausewitz, Carl von. *On War.* Edited by Michael Howard and Peter Paret, Princeton, NJ: Princeton University Press, 1976.

Cleveland, Charles T., and Daniel Egel. *The American Way of Irregular War: An Analytical Memoir.* Santa Monica, CA: RAND, 2020.

Cleveland, Charles T., Daniel Egel, Russell Howard, David Maxwell, and Hy Rothstein. "The Congressionally Authorized Irregular Warfare Functional Center Is a Historic Opportunity: Will the Department of Defense Capitalize on It?" *The RAND Blog*, November 7, 2022. https://www.rand.org/blog/2022/11/the-congressionally-authorized-irregular-warfare-functional.html.

Cleveland, Charles T., Benjamin Jensen, Susan Bryant, and Arnel David. *Military Strategy in the 21st Century: People, Connectivity, and Competition.* Amherst, NY: Cambria Press, 2018.

Cleveland, Charles T., Ryan C. Crocker, Daniel Egel, Andrew Liepman, and David Maxwell. "An American Way of Political Warfare: A Proposal." Santa Monica: RAND Corporation, 2018. https://www.rand.org/pubs/perspectives/PE304.html.

Cohen, Benjamin J. *Currency Statecraft: Monetary Rivalry and Geopolitical Ambition.* Chicago, IL: University of Chicago Press, 2018.

Cohen, Zachary. "US recently conducted cyber operation against Iran to protect election from foreign interference." *CNN,* November 3,

2020. https://www.cnn.com/2020/11/03/politics/us-cyber-operation-iran-election-interference/index.html.

Coicaud, J. M. "Deconstructing international legitimacy." In *Fault Lines of International Legitimacy*. https://doi.org/10.1111/j.1475-6765.2006.00307.x.

Colby, Elbridge A. "Testimony Before the Senate Armed Services Committee Hearing on Implementation of the National Defense Strategy." Washington, DC: Senate Armed Services Committee, January 29, 2019. https://www.armed-services.senate.gov/imo/media/doc/Colby_01-29-19.pdf.

Commission on PPBE Reform. *Commission on Planning, Programming, Budgeting, and Execution Reform: Status Update*. https://ppbereform.senate.gov/wp-content/uploads/2023/03/PPBE-REFORM-COMMISSION-STATUS-UPDATE-MAR-2023-Public.pdf;

Congressional Research Service. "China Naval Modernization: Implications for U.S. Navy Capabilities—Background and Issues for Congress." Washington, DC: Congressional Research Service, October 19, 2023. https://sgp.fas.org/crs/row/RL33153.pdf.

Conley, Heather A., James Mina, Ruslan Stefanov, and Martin Vladimirov. *The Kremlin Playbook*. Washington, DC: Center for Strategic and International Studies, 2016. https://www.csis.org/analysis/kremlin-playbook.

Connover, Pamela Johnson, Karen A. Mingst, and Lee Sigelman. "Mirror Images in Americans' Perceptions of Nations and Leaders during the Iranian Hostage Crisis." *Journal of Peace Research* 17, no. 4, (December 1980): 281–363.

Cordesman, Anthony H. "The U.S. Joint Chiefs New Strategy Paper on Joint Concept for Competing." *Center for Strategic and International Studies*, March 17, 2023. https://www.csis.org/analysis/us-joint-chiefs-new-strategy-paper-joint-concept-competing.

———, and Grace Hwang, *US Competition with China and Russia: The Crisis-Driven Need to Change U.S. Strategy*. Washington, DC: Center for Strategic and International Studies, 2020.

Correlates of War. "National Material Capabilities (v6.0)." https://correlatesofwar.org/data-sets/national-material-capabilities/.

"Costs of the 20-year war on terror: $8 trillion and 900,000 deaths." Brown University, September 1, 2021, https://www.brown.edu/news/2021-09-01/costsofwar.

Council on Foreign Relations. "China's Maritime Disputes: A CFR InfoGuide Presentation.," https://www.cfr.org/interactives/chinas-maritime-disputes?cid=otr-marketing_use-china_sea_InfoGuide#!/chinas-maritime-disputes?cid=otr-marketing_use-china_sea_InfoGuide.

Cox, Robert W., and Harold K. Jacobson. *The Anatomy of Influence: Decision Making in International Organization.* New Haven, CT: Yale University Press, 1974.

Craisor-Constantin, Ionita. "Is Hybrid Warfare Something New?" *Strategic Impact* 53, (2014): 61–71.

Crouch, Matthew, Barry Pavel, Clementine G. Starling, and Christian Trotti. "A New Strategy for US Global Defense Posture." *Strategic Insights Memo, Atlantic Council,* August 5, 2021. https://www.atlanticcouncil.org/content-series/strategic-insights-memos/a-new-strategy-for-us-global-defense-posture/.

Cronin, Patrick M. "Irregular Warfare: New Challenges for Civil-Military Relations." *Strategic Forum.* Washington, DC: National Defense University, October 2008.

CSIS, Asia Maritime Transparency Initiative. "China Island Tracker." https://amti.csis.org/island-tracker/china/.

CSIS, ChinaPower. "How Dominant Is China in the Global Arms Trade?" April 26, 2018. https://chinapower.csis.org/china-global-arms-trade/.

Cunningham, Edward, Tony Saich, and Jessie Turiel. *Understanding CCP Resilience: Surveying Chinese Public Opinion Through Time.* Cambridge, MA: Harvard Kennedy School, Ash Center for Democratic Governance and Innovation, 2020. https://ash.harvard.edu/publications/understanding-ccp-resilience-surveying-chinese-public-opinion-through-time.

Curtis, Glenn. "An Overview of Psychological Operations (PSYOP)." Washington, DC: Library of Congress, Federal Research Division, October 1989. https://apps.dtic.mil/sti/pdfs/ADA302389.pdf.

Curtis, Jesse. "Springing the 'Tacitus Trap': Countering Chinese State-Sponsored Disinformation." *Small Wars & Insurgencies* 32 (February 21, 2021): 229–265. https://doi.org/10.1080/09592318.2021.1870429.

Cyberspace Solarium Commission. "Countering Disinformation in the United States." *CSC White Paper* No. 6, (December 2021).

Dahl, Robert A. "The Concept of Power." *Behavioral Science* 2, no. 3, (1957): 201–215.

———, and Bruce Stinebrickner. *Modern Political Analysis.* Boston: Pearson, 2005.

Dalio, Ray. *Principles for Dealing with the Changing World Order.* New York, NY: Simon & Schuster, 2021.

Dar, Arshid Iqbal. "Beyond Eurocentrism: Kautilya's Realism and India's Regional Diplomacy." *Humanities & Social Sciences Communications* 8, no. 1, (December 2021): 1–7.

Defense Security Cooperation Agency. "Security Cooperation Overview." https://www.dsca.mil/foreign-customer-guide/security-cooperation-overview.

———. "Chapter 10 - International Training." https://samm.dsca.mil/chapter/chapter-10#C10.2.

———. "Global Train and Equip." https://www.dsca.mil/global-train-and-equip.

———. "International Military Training & Education Programs," https://www.dsca.mil/50th-anniversary/international-military-and-education-programs.

———. "Organization History." https://www.dsca.mil/resources/publications/strategic_plan_2025/organization_history.

———. "Section 333 Authority to Build Capacity." https://www.dsca.mil/section-333-authority-build-capacity.

Dellmuth, Lisa, and Bernd Schlipphak. "Legitimacy Beliefs towards Global Governance Institutions: A Research Agenda." *Journal of European*

Public Policy 27, no. 6 (June 2, 2020): 931–943. https://doi.org/10.10 80/13501763.2019.1604788.

Derleth, James W. "Great Power Competition, Irregular Warfare, and the Gray Zone." Irregular Warfare Center, January 13, 2023. https://irregularwarfarecenter.org/2023/01/great-power-competition-irregular-warfare-and-the-gray-zone/.

Dew, Andrea J. "Fighting for Influence in Open Societies: The Role of Resilience and Transparency." *The Fletcher Forum of World Affairs* 43, no. 2 (2019): 151–165. http://www.jstor.org/stable/45289764.

Diaz, Melissa N., Kelsi Bracmort, Phillip Brown, Richard J. Campbell, Corrie E. Clark, Mark Holt, Ashley J. Lawson, Michael Ratner, Brandon S. Tracy, and Brent D. Yacobucci. "U.S. Energy in the 21st Century: A Primer." Washington, DC: Congressional Research Service, March 16, 2021. https://crsreports.congress.gov/product/pdf/R/R46723.

Dikötter, Frank. *China After Mao: The Rise of a Superpower.* New York: Bloomsbury Publishing, 2022.

Dobbins, James, Howard J. Shatz, and Ali Wyne. "Russia Is a Rogue, Not a Peer; China Is a Peer, Not a Rogue." Santa Monica, CA: RAND, October 2018. https://www.rand.org/pubs/perspectives/PE310.html.

Donnithorne, Jeffrey W.W. *Four Guardians: A Principled Agent View of Civil Military Relations.* Baltimore: Johns Hopkins University Press, 2021.

Dostri, Omer. "The Reemergence of Gray-Zone Warfare in Modern Conflicts: Israel's Struggle against Hamas's Indirect Approach." *Military Review: The Professional Journal of the U.S. Army,* (January-February 2020): 120–127. https://www.armyupress.army.mil/Journals/Military-Review/English-Edition-Archives/January-February-2020/Dostri-Gray-Zone/.

Dowd, Anna M., and Cynthia R Cook. "Bolstering Collective Resilience in Europe." Washington, DC: Center for Strategic and International Studies, December 2022. https://www.csis.org/analysis/bolstering-collective-resilience-europe#:~:text=Rather%2C%20reflecting%20the %20North%20Atlantic,military%20threats%20to%20Euro%2DAtlantic.

Drezner, Daniel W. "Sanctions Sometimes Smart: Targeted Sanctions in Theory and Practice." *International Studies Review* 13, no. 1 (March 2011): 96–108.

———. "The United States of Sanctions: The Use and Abuse of Economic Coercion." *Foreign Affairs* 100, no. 5 (September/October 2021): 142–154.

Drinhausen, Katja, and Helena Legarda. "'Comprehensive National Security' Unleashed: How Xi's Approach Shapes China's Policies at Home and Abroad." *MERICS*, September 15, 2022. https://merics.org/en/report/comprehensive-national-security-unleashed-how-xis-approach-shapes-chinas-policies-home-and.

Drohan, Thomas A. "Bringing 'Nature of War' into Irregular Warfare Strategy: Contemporary Applications of Clausewitz's Trinity." *Defence Studies* 11, no. 3, (September 2011): 497–516. https://www.tandfonline.com/doi/abs/10.1080/14702436.2011.630176.

Drucker, Peter F. *Classic Drucker: Essential Wisdom of Peter Drucker from the Pages of Harvard Business Review.* https://search.library.wisc.edu/catalog/9910017740602121.

Dueck, Colin. *Reluctant Crusaders: Power, Culture, and Change in American Grand Strategy.* Princeton, NJ: Princeton University Press, 2008.

Dunigan, Molly, Dick Hoffman, Peter Chalk, Brian Nichiporuk, and Paul DeLuca. *Characterizing and Exploring the Implications of Maritime Irregular Warfare.* Santa Monica, CA: RAND, 2012.

Eaglen, Mackenzie. "China's Real Military Budget Is Far Bigger Than It Looks." *19fortyfive.com,* June 16, 2023. https://www.19fortyfive.com/2023/06/chinas-real-military-budget-is-far-bigger-than-it-looks/.

Eaton, Sarah, and Reza Hasmath. "Economic Legitimation in a New Era: Public Attitudes to State Ownership and Market Regulation in China." *The China Quarterly* 246 (June 2021): 447–472. https://doi.org/10.1017/S0305741020000569.

Echevarria II, Antulio J. *Reconsidering the American Way of War: US Military Practice from the Revolution to Afghanistan.* Washington, DC: Georgetown University Press, 2014.

Economy, Elizabeth C. *The Third Revolution: Xi Jinping and the New Chinese State.* New York, NY: Oxford University Press, 2018.

———. *The World According to China.* Cambridge: Polity, 2022.

Efremenko, Dmitry, Anastasia Ponamareva, and Yury Nikulichev. "Russia's Semi-Soft Power." *World Affairs: The Journal of International Issues* 25, no. 1 (2021): 100–121. https://www.jstor.org/stable/48622937.

Eitelhuber, Norbert. "The Russian Bear: Russian Strategic Culture and What It Implies for the West." *Connections: The Quarterly Journal* 9, no. 1 (Winter 2009): 1–28. https://www.proquest.com/docview/517598503/abstract/E2971980A4DC4ABFPQ/1.

Ellis, David C. "Narrative Statecraft: A Primer for Special Operations." *Kingston Consortium on International Security Insights* 2, no. 1, (January 2022): 1–6.

Energy Information Agency. "China imported record volumes of oil in 2023." September 18, 2023, https://www.eia.gov/todayinenergy/detail.php?id=60401.

Erlich, Aaron, and Calvin Garner. "Is pro-Kremlin Disinformation Effective? Evidence from Ukraine." *The International Journal of Press/Politics* 28, Issue 1, (January 2023): 5–28.

Eshel, Yohanan, Shaul Kimhi, and Hadas Marciano. "Predictors of National and Community Resilience of Israeli Border Inhabitants Threatened by War and Terror." *Community Mental Health Journal* 56, no. 8, (November 2020): 1480–1488.

Fabian, Sandor. "US IMET Participation and the Duration and Outcome of Insurgencies." *Defense Studies,* 21, no. 2, (2021): 242–265.

Fabian, Sandor, and Gabrielle Kennedy. "Divided by a Common Language: How Europe Views Irregular Warfare." *Irregular Warfare Initiative,* May 31, 2023. https://irregularwarfare.org/articles/divided-by-a-common-language-how-europe-views-irregular-warfare/.

Farrell, Henry, and Abraham L. Newman. "Weaponized Interdependence: How Global Economic Networks Shape State Coercion." *International Security* 44, no. 1 (2019): 42–79.

Farrell, Theo. "Strategic Culture and American Empire." *The SAIS Review of International Affairs* 25, no. 2, (2005): 3–18.

FBI. "Combating Foreign Influence" https://www.fbi.gov/investigate/counterintelligence/foreign-influence.

Fearon, James D. "Signaling Foreign Policy Interests: Tying Hands Versus Sinking Costs." *Journal of Conflict Resolution,* 41, no. 1 (1997): 68–90.

Federal Emergency Management Agency (FEMA). "Biden-Harris Administration Announces Nearly $2 Billion in Available Funding to Increase Resilience Nationwide." October 12, 2023. https://www.fema.gov/press-release/20231012/biden-harris-administration-announces-nearly-2-billion-available-funding.

———. *National Response Framework.* Washington, DC: U.S. Department of Homeland Security, January 2008. https://www.fema.gov/pdf/emergency/nrf/nrf-core.pdf.

———. *National Response Framework* (Fourth Edition). Washington, DC: U.S. Department of Homeland Security. October 28, 2019. https://www.fema.gov/sites/default/files/2020-04/NRF_FINALApproved_20 11028.pdf.

Fefer, Rachel F. "Section 232 of the Trade Expansion Act of 1962." Washington, DC: Congressional Research Service, April 1, 2022. https://crsreports.congress.gov/product/pdf/IF/IF10667.

———, Keigh E. Hammond, Vivian C. Jones, Brandon J. Murrill, Michaela D. Platzer, and Brock R. Williams. "Section 232 Investigations: Overview and Issues for Congress." Washington, DC: Congressional Research Service, May 18, 2021. https://sgp.fas.org/crs/misc/R45249.pdf.

Felbermayr, Gabriel, Yoto V. Yotov, and Erdal Yalcin. "The World Trade Organization at 25: Assessing the Economic Value of the Rules Based Global Trading System." Policy report prepared for Bertelsmann Stiftung, December 30, 2019.

Fiala, Otto C, Kirk Smith, and Anders Löfberg. *Resistance Operating Concept (ROC).* Tampa: Joint Special Operations University Press, 2020. https://jsou.edu/Press/PublicationDashboard/25.

Filloux, Frederic. "You can't sell news for what it costs to make." *The Walkley Magazine on Medium*, August 6, 2017.

Finley, Sonya. "A 3Cs Framework for Conceptualizing the Competitive Information Environment." In *Teaching Public Diplomacy and The Information Instruments of Power in a Complex Media Environment*, edited by Vivian S. Walker and Sonya Finley, 6-10. Washington DC: Advisory Commission on Public Diplomacy, 2020.

Finnemore, Martha, and Judith Goldstein. *Back to Basics: State Power in a Contemporary World.* Oxford: Oxford University Press, 2013.

Fischer, Beth A. "Building Up and Seeking Peace: President Reagan's Cold War Legacy." In *Reagan's Legacy in a World Transformed*, edited by Jeffrey L. Chidester and Paul Kengor, 165-77. Cambridge, MA: Harvard University Press, 2015.

Fischer, Joseph R. "The U.S. Army and Irregular Warfare: 1775–2007." *The Journal of Military History* 74, no. 3, (2010): 917-919.

Foreign Intelligence Surveillance Act of 1978, Public Law 95-511, 95th Cong., 2d sess. (Government Printing Office, 1978). https://www.gpo.gov/fdsys/pkg/STATUTE-92/pdf/STATUTE-92-Pg1783.pdf.

Foreign Malign Influence Center, 50 U.S.C. § 3059 (e)(2), (2022). https://www.law.cornell.edu/uscode/text/50/3059#e_2.

Fracchia, Joseph, and R.C. Lewontin. "Does Culture Evolve?" *History and Theory* 38, no. 4, (December 1999): 52–78.

Fravel, M. Taylor. "China's Sovereignty Obsession." *Foreign Affairs*, June 26, 2020. https://www.foreignaffairs.com/articles/china/2020-06-26/chinas-sovereignty-obsession.

Freedman, Lawrence. "Strategy: The History of an Idea." In *The New Makers of Modern Strategy: From the Ancient World to the Digital Age*, edited by Hal Brands, 17–40. Princeton NJ: Princeton University Press, 2023.

French, John R. P., Jr. "A Formal Theory of Social Power." *Psychological Review* 63, (1956): 181–194.

Freedom House. "Freedom in the World." https://freedomhouse.org/report/freedom-world.

Bibliography 255

Friedberg, Aaron L. "Globalisation and Chinese Grand Strategy." *Survival* 60, no. 1 (January 2, 2018): 7–40. https://doi.org/10.1080/00396338.2 018.1427362.

Friedland, Nehemia. "Introduction – The elusive concept of social resilience." In *The Concept of Social Resilience*, edited by Nehemia Friedland, Asher Arian, Alan Kirschenbaum, Karin Amit, and Nicole Fleischer, 1-15. Tel Aviv: Samuel Neaman Institute, 2005.

———, Asher Arian, Alan Kirschenbaum, Karin Amit, and Nicole Fleischer, eds. *The Concept of Social Resilience*. Tel Aviv: Samuel Neaman Institute, 2005.

Friedman, George. "Still A Unipolar World." *Geopolitical Futures*, October 18, 2022. https://geopoliticalfutures.com/still-a-unipolar-world/.

Friedman, Uri. "The Pandemic Is Revealing a New Form of National Power." *The Atlantic*, November 15, 2020. https://www.theatlantic.com/ideas/archive/2020/11/pandemic-revealing-new-form-national-power/616944/.

Fukuyama, Francis. "The Thing That Determines a Country's Resistance to the Coronavirus." *The Atlantic*, March 30, 2020. https://www.theatlantic.com/ideas/archive/2020/03/thing-determines-how-well-countries-respond-coronavirus/609025/.

Fulton, Jonathan, ed. *Routledge Handbook on China–Middle East Relations*. London: Routledge, 2021.

Fund for Peace. "State Resilience Index." https://www.fundforpeace.org/SRI/about.html.

———. "State Resilience Index Annual Report 2022." December 7, 2022, https://tgpcloud.org/modules/blogs/blogDisplay.php?r=84.

———. "State Resilience Index: Measuring Capacities and Capabilities In 154 Countries." https://www.fundforpeace.org/SRI/about.html.

Gaddis, John Lewis. *Strategies of Containment*. London: Oxford University Press, 2005.

———. "Was the Truman Doctrine a Real Turning Point?" *Foreign Affairs* 52, no. 2 (January 1974): 386–402.

Galeotti, Mark. "Active Measures: Russia's Covert Geopolitical Operations." *Security Insights, George C. Marshall Center for European Security*, Number 31, June 2019. https://www.marshallcenter.org/en/publications/security-insights/active-measures-russias-covert-geopolitical-operations-0

———. *The Weaponisation of Everything*. Audible Studios, 2022. (Audiobook). https://www.amazon.com/Weaponisation-Everything-Field-Guide-New/dp/B09RQ29SLN/ref=sr_1_1?crid=2QSFZP6U3FIGA&keywords=the+weaponization+of+everything&qid=1680805259&sprefix=the+weaponizat%2Caps%2C81&sr=8-1.

Gallup Polls. "China." In *Depth Topics: A to Z* (blog), October 16, 2023. https://news.gallup.com/poll/1627/china.aspx.

Galtung, Jonathan. "An Editorial." *Journal of Peace Research* 1, no. 1 (1964): 2.

Garamone, Jim. "Noncommissioned Officers Give Big Advantage to U.S. Military." U.S. Department of Defense, November 7, 2019. https://www.defense.gov/News/News-Stories/article/article/2011393/noncommissioned-officers-give-big-advantage-to-us-military/.

Gartzke, Erik, and Oliver Westerwinter. "The complex structure of commercial peace contrasting trade interdependence, asymmetry, and multipolarity." *Journal of Peace Research* 53, no. 3, (May 2016): 325–343. https://doi-org.proxy.library.georgetown.edu/10.1177/0022343316637895.

Gazis, Olivia. "Bipartisan Senate Intel Report Backs Intelligence Assessment of 2016 Russian Interference." *CBS News*, April 21, 2020. https://www.cbsnews.com/news/senate-intelligence-committee-report-2016-russian-interference-assessment/.

George Washington's Mount Vernon. "Washington's Farewell Address, 1796." https://www.mountvernon.org/education/primary-source-collections/primary-source-collections/article/washington-s-farewell-address-1796/.

Gartner, Scott Sigmund, and Marissa Edson Myers. "Body Counts and 'Success' in the Vietnam and Korean Wars." *The Journal of*

Interdisciplinary History 25, no. 3 (1995): 377–395. https://doi.org/1 0.2307/205692.

George, Roger Z., and Harvey Rishikof. *The National Security Enterprise: Navigating the Labyrinth.* Washington, DC: Georgetown University Press, 2010.

German, Tracey. "Harnessing Protest Potential: Russian Strategic Culture and the Colored Revolutions." *Contemporary Security Policy* 41, no. 4, (October 1, 2020): 541–563.

———, and Natasha Kuhrt. "Will Russia's Global Influence Continue to Decline?" *King's College London*, February 21, 2023. https://www.kcl. ac.uk/will-russias-global-influence-continue-to-decline.

Gerschewski, Johannes. "The Three Pillars of Stability: Legitimation, Repression, and Co-Optation in Autocratic Regimes." *Democratization* 20, no.1 (2013): 13–38 DOI: 10.1080/13510347.2013.738860.

Gershaneck, Kerry K. *Political Warfare: Strategies for Combating China's Plan to "Win Without Fighting."* Quantico, VA: Marine Corps University Press, 2020.

Gilchrist, Mark. "Why Thucydides Still Matters." *The Strategy Bridge*, November 30, 2016. https://thestrategybridge.org/the-bridge/2016/1 1/30/why-thucydides-still-matters.

Gilley, Bruce. "The Meaning and Measure of State Legitimacy: Results for 72 Countries." *European Journal of Political Research* 45, no. 3 (2006): 499–525. https://doi.org/10.1111/j.1475-6765.2006.00307.x.

Giske, Mathilde Tomine Eriksdatter. "Resilience in the Age of Crises." Norwegian Institute of International Affairs, 2021. https://www.jstor. org/stable/resrep30682.

Gleiman, J.K. *The Unconventional Strategic Option.* Kansas State University, 2018. https://krex.k-state.edu/handle/2097/38932.

Goldenziel, Jill I., and Manal Cheema. "The New Fighting Words?: How U.S. Law Hampers the Fight Against Information Warfare." 22 U. Pa. J. Const. L. 81. February 26, 2019. https://ssrn.com/abstract=3286847.

Goodhart, Charles. "Problems of monetary management: the UK experience in papers in monetary economics." *Monetary Economics* 1 (1975).

Gopnik, Adam. "How Samuel Adams Helped Ferment A Revolution." *The New Yorker*, October 24, 2022. https://www.newyorker.com/magazine/2022/10/31/how-samuel-adams-helped-ferment-a-revolution.

Gordon, Joy. "The Not So Targeted Instrument of Asset Freezes." *Ethics & International Affairs* 33, no. 3 (2019): 303–314. https://doi.org/10.1017/S0892679419000315.

Gordon, Vikki. "The Law: Unilaterally Shaping U.S. Security Policy: The Role of National Security Directives." *Presidential Studies Quarterly* 37, no. 2 (June 2007): 349–367.

Gorny, Thorsten J. "Primary and Secondary Sanctions Explained." *Sanctions.io*, June 28, 2022. https://www.sanctions.io/blog/primary-and-secondary-sanctions-explained.

Graham, Thomas. "The Precarious Future of Russian Democracy." *Council on Foreign Relations*, March 7, 2023. https://www.cfr.org/article/precarious-future-russian-democracy;

Grandin, Greg. "Off the Beach: The United States, Latin America, and the Cold War." In *A Companion to Post-1945 America*, edited by J. Agnew and R. Rosenzweig 426-445. Newark NJ: John Wiley & Sons, 2002.

Grasso, Valerie Bailey. "Defense Acquisition Reform: Status and Current Issues." Washington, DC: Congressional Research Service, September 7, 2000. https://apps.dtic.mil/sti/pdfs/ADA481142.pdf.

Gray, Colin S. *Categorical Confusion? The Strategic Implications of Recognizing Challenges Either as Irregular or Traditional.* Carlisle, PA: Strategic Studies Institute, U.S. Army War College, 2012.

———. *Geopolitics and Strategic Culture.* Lexington, KY: University of Kentucky Press, 1988.

———. *The Geopolitics of Super Power.* Lexington, KY: University of Kentucky Press, 1989.

———. "Irregular Enemies and the Essence of Strategy: Can the American Way of War Adapt?" Carlisle, PA: Strategic Studies Institute, US Army War College, 2006.

———. "Irregular Warfare: One Nature, Many Characters." *Strategic Studies Quarterly* 1, no. 2, (Winter 2007): 35–57.

———. *Modern Strategy.* Oxford, UK: Oxford University Press, 1999.

———. "National Style in Strategy: The American Example." *International Security* 6, no. 2 (1981): 21-47, https://www.jstor.org/stable/2538645?origin=crossref.

———. "Out of the Wilderness: Prime Time for Strategic Culture." Washington, DC: Defense Threat Reduction Agency, October 31, 2006. https://irp.fas.org/agency/dod/dtra/stratcult-out.pdf.

———. "Strategic Culture as Context: The First Generation Theory Strikes Back." *Review of International Studies* 25, no. 1 (January 1999): 49-69.

———. "Why Strategy is Difficult." *Joint Force Quarterly* 22, (Summer 1999): 6-12.

Green, Mark A. "Russian Arms Exports." *The Wilson Center,* May 17, 2022. https://www.wilsoncenter.org/blog-post/russian-weapon-exports.

Greenhill, Jim. "State Partnership Program Turns 30: A Crucial Arrow in Ukraine's Quiver." U.S. Department of Defense, July 17, 2023. https://www.defense.gov/News/News-Stories/Article/Article/3460799/state-partnership-program-turns-30-a-crucial-arrow-in-ukraines-quiver/.

Greenwalt, William. *Leveraging the National Technology Industrial Base to Address Great-Power Competition.* Washington, DC: Atlantic Council, Scowcroft Center for Strategy and Security, April 2019. https://www.atlanticcouncil.org/wp-content/uploads/2019/04/Leveraging_the_National_Technology_Industrial_Base_to_Address_Great-Power_Competition.pdf.

Greitens, Sheena Chestnut. "Xi's Security Obsession." *Foreign Affairs,* July 28, 2023. https://www.foreignaffairs.com/china/xis-security-obsession.

Grisé, Michelle, Alyssa Demus, Yuliya Shokh, Marta Kepe, Jonathan W. Welburn, and Khrystyna Holynska. *Rivalry in the Information Sphere.* Santa Monica, CA: RAND, 2022. https://doi.org/10.7249/RRA198-8.

Griswold, Daniel. "Four Decades of Failure: The U.S. Embargo against Cuba." CATO Institute, May 12, 2005. https://www.cato.org/speeches/four-decades-failure-us-embargo-against-cuba.

Grundman, Steven C. *The Monopsonist's Dilemma*. Boston, MA: Charles River Associates, October 2010. http://grundmanadvisory.com/wp-content/uploads/2010/10/CRA-Insights-on-Monopsonists-Dilemma-Oct-2010.pdf.

Gui, Yongtao. "Moving toward Decoupling and Collective Resilience? Assessing US and Japan's Economic Statecraft against China." *China International Strategy Review* 4, no. 1, (June 2022): 55–73. https://doi.org/10.1007/s42533-022-00097-z.

Gupta, Amit. "Panda Power? Chinese Soft Power in the Era of COVID-19." *PRISM* 10, no. 1 (2022): 40–56. https://www.jstor.org/stable/48697206.

Guriev, Sergei, and Daniel Treisman. *Spin Dictators: The Changing Face of Tyranny in the 21st Century*. Princeton, NJ: Princeton University Press, 2022. https://doi.org/10.2307/j.ctv1xp9p7d.

Gustafson, Michael, and James M. Nyce. "Modern Society, Violence and Irregular Warfare: A Culturally Informed Critique." *Diskussion & Debatt* 1, (January/March 2012): 136–150.

Hadley, Stephen J., Jason Miler, and Mara E. Carlin. "Connecting Stategy and Resources In National Security." Washington, DC: The Atlantic Council, 2019.

Haenle, Paul. "China's Zero COVID Policy is a Double-Edged Sword." Carnegie Endowment for International Peace, December 1, 2022. https://carnegieendowment.org/2022/12/01/china-s-zero-covid-policy-is-double-edged-sword-pub-88535.

Haken, Nate, Daniel Woodburn, Emily Sample, Wendy Wilson, John Madden, et al. "Fragile States Index 2023 – Annual Report," June 14, 2023. https://fragilestatesindex.org/2023/06/14/fragile-states-index-2023-annual-report/.

Heldenwang, Christian von. "Measuring Legitimacy--New Trends, Old Shortcomings?" *German Development Institute* 18, (2016).

Hall, Holly Kathleen. "The New Voice of America: Countering Foreign Propaganda and Disinformation Act." *First Amendment Studies* 51, no. 2 (July 3, 2017): 49–61. https://doi.org/10.1080/21689725.2017.1349618.

Bibliography

Hall, Stephen G.F. and Thomas Ambrosio. "Authoritarian learning: a conceptual overview." *East European Political* 33, Issue 2, (May 5, 2017): 143–161.

Hammes, T.X. *An Affordable Defense of Asia*. Washington, DC: Atlantic Council, June 2022. https://www.atlanticcouncil.org/wp-content/uploads/2020/06/An-Affordable-Defense-of-Asia-Report.pdf.

Hancock, Larry. *Creating Chaos: Covert Political Warfare, from Truman to Putin*. New York, NY: OR Books, 2018.

Harrell, Peter, Elizabeth Rosenberg, and Edoardo Saravalle. "China's Use of Coercive Economic Measures," Washington, DC: Center for a New American Security, 2018. http://www.jstor.org/stable/resrep20429.6.

Harold, Scott W., Nathan Beauchamp-Mustafaga, and Jeffrey W. Hornung. *Chinese Disinformation Efforts on Social Media*. Santa Monica, CA: RAND, 2021, https://www.rand.org/pubs/research_reports/RR4373z3.html.

Hasselbach, Christoph. "SIPRI: US Arms Exports Skyrocket, While China's Nosedive." *Deutsche Welle* (blog), March 13, 2023. https://www.dw.com/en/sipri-us-arms-exports-skyrocket-while-chinas-nosedive/a-64948062.

Health Canada. "Public Health Update on the Health Effects of Wildfires." *Government of Canada*, June 19, 2023. https://www.canada.ca/en/health-canada/news/2023/06/public-health-update-on-the-health-effects-of-wildfires.html.

Heike, Paul. *The Myths That Made America*. Bielefeld, DE: Transcript-Verlag, 2014.

Heil, Alan L., Jr. *Voice of America*. New York, NY: Columbia University Press, 2003.

Helmus, Todd C., and Marta Kepe. "A Compendium of Recommendations for Countering Russian and Other State-Sponsored Propaganda." Santa Monica, CA, RAND Corporation, June 1, 2021. https://www.rand.org/pubs/research_reports/RRA894-1.html.

Herd, Graeme P. *Understanding Russian Strategic Behavior: Imperial Strategic Culture and Putin's Operational Code.* London: Routledge, 2022.

———. Ed. *Russia's Global Reach: A Security and Statecraft Assessment.* Garmisch-Partenkirchen: George C. Marshall Center for European Security Studies, 2021.

Herman, Arthur. *Freedom's Forge: How American Business Produced Victory in World War II,* New York, NY: Random House, 2013.

Hicks, Kathleen, Alice Hunt Friend, Joseph P. Federici, Hijab Shah, Megan Donahoe, Matthew Conklin, Asya Akca, Michael Matlaga, and Lindsey Sheppard. "By Other Means Part I: Campaigning in the Gray Zone." Washington, DC: Center for Strategic and International Studies, July 8, 2019.

Himmer, Michal, and Zdeněk Rod. "Chinese Debt Trap Diplomacy: Reality or Myth?" *Journal of the Indian Ocean Region* 18, no. 3 (September 2, 2022): 250–272. https://doi.org/10.1080/19480881.2023.2195280.

Hitchens, Harold L. "Influences on the Congressional Decision to Pass the Marshall Plan." *The Western Political Quarterly* 21, no. 1 (March 1968): 51–68.

Hlavka, Jakub and Adam Rose. "Unprecedented by most measures: Calculating the astonishing costs of COVID." *Los Angeles Times,* May 15, 2023. https://www.latimes.com/opinion/story/2023-05-15/covid-pandemic-us-economic-costs-14-trillion.

Hodicky, Jan, Gökhan Özkan, Hilmi Özdemir, Petr Stodola, Jan Drozd, and Wayne Buck. "Dynamic Modeling for Resilience Measurement: NATO Resilience Decision Support Model." *Applied Sciences* 10, no. 8 (January 2020): 2639. https://doi.org/10.3390/app10082639.

Hoffman, Frank. *Conflict in the 21st Century: The Rise of Hybrid Wars.* Arlington, VA: Potomac Institute for Policy Studies, 2007.

———. "Examining Complex Forms of Conflict: Gray Zone and Hybrid Challenges." *PRISM* 7, no. 4, (November 8, 2018): 30–47.

———. "Irregular Warfare: One Nature, Many Characters." *Strategic Studies Quarterly* 1, no. 2, (Winter 2007): 35–57.

———. "On Not-So-New Warfare: Political Warfare vs Hybrid Threats." *War on the Rocks*, July 28, 2014. https://warontherocks.com/2014/07/on-not-so-new-warfare-political-warfare-vs-hybrid-threats/.

———. "Review: Strategic Culture And Ways Of War: Elusive Fiction Or Essential Concept?" *Naval War College Review* 70, No. 2 (2017): 137–144. https://www.jstor.org/stable/26398024?pq-origsite=summon.

Howard, Philip N., Bharath Ganesh, Dimitra Liotsiou, John Kelly, and Camille François. "The IRA, Social Media and Political Polarization in the United States, 2012–2018." U.S. Senate Documents, October 2019.

Howard, Russell D. *Strategic Culture*. Macdill Air Force Base, FL: The JSOU Press, 2013.

Hua, Shipping, and Ka Zeng. "The US-China Trade War: Economic Statecraft, Multinational Corporations and Public Opinion." *Business and Politics* 24 (2022): 319–331.

Huang, Yasheng. *The Rise and Fall of the East*. New Haven, CT: Yale University Press, 2023.

Hudson, Valerie M. *Culture and Foreign Policy*. London: Reinner Publishers, 1997.

Hufbauer, Gary. *Economics Sanctions Reconsidered*. Washington, DC: Peterson Institute for International Economics, 2009.

Hunt, Jonathan R. and Simon Miles. *The Reagan Moment: America and the World in the 1980s*. Ithaca, NY: Cornell University Press, 2021.

Huntington, Samuel. *The Soldier and the State*. Cambridge, MA: Belknap Press, 1956.

Hurd, Ian. "Legitimacy." *The Princeton Encyclopedia of Self-Determination*. https://pesd.princeton.edu/node/516.

Hurley, W.J., C.R. Bucher, S.K. Numrich, S.M. Ouellette, and J.B. Resnick. "Non-Kinetic Capabilities for Irregular Warfare: Four Case Studies." Alexandria, VA: Institute for Defense Analysis, March 2009.

Ignatieff, Michael. *Virtual War: Kosovo and Beyond*. 1st Edition, New York, NY: Picador USA, 2000.

———, ed. *American Exceptionalism and Human Rights.* Princeton, NJ: Princeton University Press, 2005.

Insinna, Valerie. "Biden Administration Kills Trump-Era Nuclear Cruise Missile Program." *Breaking Defense* (blog), March 28, 2022. https://breakingdefense.sites.breakingmedia.com/2022/03/biden-administration-kills-trump-era-nuclear-cruise-missile-program/.

International Crisis Group. "Competing Visions of International Order in the South China Sea." Report No. 315, November 29, 2021. https://www.crisisgroup.org/asia/north-east-asia/china/315-competing-visions-international-order-south-china-sea.

———. "Sanctions, Peacemaking and Reform: Recommendations for U.S. Policymakers." Brussel, Belgium: United States Report N°8, August 28, 2023. https://www.crisisgroup.org/united-states/8-sanctions-peacemaking-and-reform-recommendations-us-policymakers.

International Federation of Red Cross and Red Crescent Societies. "IFRC Framework for Community Resilience." 2014. https://www.ifrc.org/sites/default/files/IFRC-Framework-for-Community-Resilience-EN-LR.pdf.

International Monetary Fund, "Currency Composition of Official Foreign Exchange Reserves (COFER)." https://data.imf.org/?sk=e6a5f467-c14b-4aa8-9f6d-5a09ec4e62a4.

"International Reserves of the Russian Federation (End of period)." *Databases, Bank of Russia*, last updated September 15, 2023. https://www.cbr.ru/eng/hd_base/mrrf/mrrf_7d/.

Jack, Caroline. "Lexicon of Lies: Terms for Problematic Information." *Data&Society*, August 9, 2017. https://datasociety.net/library/lexicon-of-lies/.

Jacobs, Charity S., and Kathleen M. Carley. "#WhatIsDemocracy: Finding Key Actors in a Chinese Influence Campaign." *Computational and Mathematical Organization Theory*, July 1, 2023. https://doi.org/10.1007/s10588-023-09380-9.

Jacobsen, Carl G., ed. *Strategic Power: The United States of America and the U.S.S.R.* Palgrave MacMillan, 1990.

Jacobsen, Katherine. "In 2020, U.S. Journalists Faced Unprecedented Attacks," *Committee to Protect Journalists*, December 14, 2020, https://cpj.org/2020/12/in-2020-u-s-journalists-faced-unprecedented-attacks/#:~:text=At%20least%20110%20journalists%20were,CPJ%20is%20a%20founding%20member.

Jensen, Rebecca, and Larry Caswell. "Deterrence by Resilience: The Case of Ukraine." *Turkish Policy*, June 19, 2023. http://turkishpolicy.com/article/1204/deterrence-by-resilience-the-case-of-ukraine.

Jin, Emily. "Why China's CIPS Matters (and Not for the Reasons You Think)." *Lawfare* (blog), April 5, 2022. https://www.lawfaremedia.org/article/why-chinas-cips-matters-and-not-reasons-you-think.

Jin, Keyu. *The New China Playbook*. New York: Viking, 2023.

Johnson, Cathryn, Timothy J. Dowd, and Cecilia L. Ridgeway. "Legitimacy as a Social Process." *Annual Review of Sociology* 32 (2006): 53–78. https://www.jstor.org/stable/29737731.

Johnson, David E. *Hard Fighting: Israel in Lebanon and Gaza*. Santa Monica: CA, RAND, 2012.

Johnson, Jeannie L. "Fit for Future Conflict? American Strategic Culture in the Context of Great Power Competition." *Marine Corps University Journal* 11, no. 1 (2020): 185–208. https://www.usmcu.edu/Portals/218/JAMS_11_1_Fit%20for%20Future%20Conflict_Jeannie%20Johnson.pdf.

Johnson, Jeannie L. "The 'Know Thyself' Conundrum: Extracting Strategic Utility from a Study of American Strategic Culture." In *Strategy Matters*, edited by Donovan C. Chau, 113-136. Maxwell Air Base AL: Air University Press, 2022.

Johnson, Keith. "Putin's Gas Gambit Backfires." *Foreign Policy*, December 12, 2013. https://foreignpolicy.com/2013/12/12/putins-gas-gambit-backfires/

Johnson, Robert. "Hybrid War and Its Countermeasures: A Critique of the Literature." *Small Wars & Insurgencies* 29, no. 1, (2018): 141–163.

Johnston, Alastair Iain. *Cultural Realism: Strategic Culture and Grand Strategy in Chinese History.* Princeton, NJ: Princeton University Press, 2020. (Kindle Book).

———. "Thinking about Strategic Culture." *International Security* 19, no. 4 (Spring 1995): 32–64.

Joint Chiefs of Staff. *Information in Joint Operations* (JP 3-04), 2022.

Joint Staff. "Joint Publication 3-22: Foreign Internal Defense."

Joint Publication. *Information Operations.* https://defenseinnovationmarketplace.dtic.mil/wp-content/uploads/2018/02/12102012_io1.pdf.

Jones, Seth G. "The Future of Competition: U.S. Adversaries and the Growth of Irregular Warfare." *Center for Strategic & International Studies*, February 4, 2021. https://www.csis.org/analysis/future-competition-us-adversaries-and-growth-irregular-warfare.

———. "Going on the Offensive: A U.S. Strategy to Combat Russian Information Warfare." *CSIS*, October 1, 2018. https://www.csis.org/analysis/going-offensive-us-strategy-combat-russian-information-warfare.

———. "The Return of Political Warfare." (Washington, DC: Center for Strategic and International Studies, February 2, 2018). https://www.csis.org/analysis/return-political-warfare.

———. *Three Dangerous Men.* New York, NY: W. W. Norton & Company, 2021.

Jordan, Amos A., William J. Taylor, Jr., Michael J. Meese, and Suzanne C. Nielsen. *American National Security.* Baltimore, MD: Johns Hopkins University Press, 2009.

Kamal, Kajari. "Kautilya's Arthashastra: Indian Strategic Culture and Grand Strategic Preferences." *Journal of Defence Studies* 12, no. 2, (July–September 2018): 27–54.

———. *Kautilya's Arthashastra: Strategic Cultural Roots of India's Contemporary Statecraft.* New York: Routledge, 2023.

Kamarick, Elaine, and Jordan Muchnik. "One Year into the Ukraine War-What Does the Public Think about American Involvement in the World?" The Brookings Institution, February 23, 2023. https://www.brookings.edu/articles/one-year-into-the-ukraine-war-what-does-the-public-think-about-american-involvement-in-the-world/.

Kanet, Roger E. "Russian Strategic Culture, Domestic Politics and Cold War 2.0." *European Politics & Society* 20, no. 2, (April 2019): 190–206.

Kapantai, E., Christopoulou, A., Berberidis, C., and Peristeras, V. "A systematic literature review on disinformation: Toward a unified taxonomical framework." *New Media & Society* 23, no. 5, (2021): 1301–1326. https://doi.org/10.1177/1461444820959296.

Kari, Martti J., and Katri Pynnöniemi. "Theory of Strategic Culture: An Analytical Framework for Russian Cyber Threat Perception." *Journal of Strategic Studies* 46, no. 1, (2023): 56–84.

"KATUSA Soldier Program." *Eighth Army*. http://8tharmy.korea.army.mil/site/about/katusa-soldier-program.asp.

Katz, David J. "Waging Financial Warfare: Why and How." *The US Army War College Quarterly: Parameters* 47, no. 2 (June 1, 2017). https://doi.org/10.55540/0031-1723.2930.

Katz, Eric. "Unfrozen; How the State Department has Reversed its 'Draconian' Cuts in Just Two Years." *Government Executive*, March 7, 2023. https://www.govexec.com/workforce/2023/03/unfrozen-how-state-department-reversed-draconian-cuts/383508/.

Kauppi, Mark V. "Thucydides: Character and capabilities." *Security Studies* 5, no. 2, (Winter 1996): 142–168.

Kautilya. *Kautilya's Arthashastra*. Translated by Rudrapatna Shamasastry. Bangalore: Sofia Publisher, 2022.

Kavanaugh, Dennis. *Political Culture*. London: MacMillan, 1972.

Kavanagh, Jennifer, and Michael D. Rich. *Truth Decay*. Santa Monica, CA: RAND, 2018. https://www.rand.org/pubs/research_reports/RR2314.html.

Kearney, Michael. "The Prohibition of Propaganda for War in the International Covenant on Civil and Political Rights." *Netherlands Quarterly of Human Rights* 23, issue 4, (December 2005): 551–570. https://Doi.org/10.1177/016934410502300402.

Keck, Marcus, and Patrick Sakdalpolrak. "What Is Social Resilience? Lessons Learned and Ways Forward." *Erkunde* 67, no. 1 (March 2013): 5–19.

Kennan, George F. "Foreign Relations of the United States, 1945–1950, Emergence of the Intelligence Establishment - Office of the Historian" (Records of the National Security Council | RG 273, Box 11, May 4, 1948). https://history.state.gov/historicaldocuments/frus1945-50Intel/d269.

———. "Policy Planning Staff Memorandum," *Office of the Historian, U.S. Department of State*, May 4, 1948. https://history.state.gov/historicaldocuments/frus1945-50Intel/d269.

Kennedy, Paul. *The Rise and Fall of the Great Powers: Economic Change and Military Conflict from 1500 to 2000*. New York, NY: Vintage, 1989.

Kennedy, Scott, Logan Wright, John L. Holden and Claire Reade. "Experts React: China's Economic Slowdown: Causes and Implications." *CSIS*, August 30, 2023. https://www.csis.org/analysis/experts-react-chinas-economic-slowdown-causes-and-implications.

Kerrane, Evan. "Moscow's Strategic Culture: Russian Militarism in an Era of Great Power Competition." *Journal of Advanced Military Studies* 2022, no. Special (2022): 69–87. https://doi.org/10.21140/mcuj.2022 SIstratcul005.

Khalid, Asma. "Biden kept Trump's tariffs on Chinese imports. This is who pays the price." *NPR Politics*, podcast audio, June 27, 2023. https://www.npr.org/2023/06/27/1184027892/china-tariffs-biden-trump.

Khanapurkar, Uday. "CFIUS 2.0: An Instrument of American Economic Statecraft Targeting China." *Journal of Current Chinese Affairs* 48, no. 2, (2020): 226–240.

Kim, Jo. "Why China's Propaganda Efforts So Often Backfire." *The Diplomat*, April 24, 2020. https://thediplomat.com/2020/04/why-chinas-propaganda-efforts-so-often-backfire/;

Kim, Patricia M., and Mallie Prytherch. "Douzheng: Unraveling Xi Jinping's Call for 'Struggle.'" *Brookings* (blog), November 2, 2023. https://www.brookings.edu/articles/douzheng-unraveling-xi-jinpings-call-for-struggle/.

Kimhi, Shaul, and Yohanan Eshel. "Individual and Public Resilience and Coping With Long-Term Outcomes of War." *Journal of Applied Biobehavioral Research* 14, no. 2 (2009): 70–89. https://doi.org/10.11 11/j.1751-9861.2009.00041.x.

———, Dmitry Leykin, and Mooli Lahad. "Individual, Community, and National Resilience in Peace Time and in the Face of Terror: A Longitudinal Study." *Journal of Loss and Trauma* 22, no. 8, (November 17, 2017): 698–713. https://doi.org/10.1080/15325024.2017.1391943.

———. "Measuring national resilience: A new short version of the scale (NR-13)." *Journal of Community Psychology* 47, (October 8, 2018).

———. "National Resilience: A New Self-Report Assessment Scale." *Community Mental Health Journal* 55, no. 4 (May 1, 2019): 721–731. https://doi.org/10.1007/s10597-018-0362-5.

———. "A New Perspective on National Resilience: Components and Demographic Predictors." *Journal of Community Psychology* 44, no. 7, (2016): 833–844.

Kimhi, Shaul, Hadas Marciano, Yohanan Eshel, and Bruria Adini. "Community and National Resilience and Their Predictors in Face of Terror." *International Journal of Disaster Risk Reduction* 50 (November 2020): 101746. https://doi.org/10.1016/j.ijdrr.2020.101746.

———. "Do we cope similarly with different adversities? COVID-19 versus armed conflict." *BMC Public Health* 22, no. 1, (November 23, 2022): 2151. https://doi.org/10.1186/s12889-022-14572-0.

Kissinger, Henry. *On China.* Reprint. New York, NY: Penguin Books, 2012.

Klare, Michael T. *All Hell Breaking Loose: The Pentagon's Perspective on Climate Change.* New York, NY: Metropolitan Books, 2019.

Klein, Hans. "Information Warfare and Information Operations: Russian and U.S. Perspectives." *Journal of International Affairs* 71, no. 1.5 (2018): 135–142.

Kliman, Daniel, Andrea Kendall-Taylor, Kristine Lee, Joshua Fitt, and Carisa Nietsche. *Dangerous Synergies: Countering Chinese and Russian*

Digital Influence Operations. Washington, DC: Center for a New American Security, 2020. http://www.jstor.org/stable/resrep25314.8.

Knock, Thomas. *To End All Wars: Woodrow Wilson and the Quest for a New World Order*. Princeton, NJ: Princeton University Press, 1995.

Knorr, Klaus. *The Power of Nations: The Political Economy of International Relations*. New York, NY: Basic Books, 1975.

Koh, Harold. "America's Jekyll and Hyde Exceptionalism." In *American Exceptionalism and Human Rights*, edited by Michael Ignatieff, 111-143. Princeton NJ: Princeton University Press, 2005.

Krause, Keith. "Military Statecraft: Power and Influence in Soviet and American Arms Transfer Relationships." *International Studies Quarterly* 35, no. 3 (1991): 313–336. https://doi.org/10.2307/2600702.

Kreps, Sarah E. "The 2006 Lebanon War: Lessons Learned." *The US Army War College Quarterly: Parameters* 37, no. 1, (Spring 2007): 72–84.

Krieg, Andreas, and Jean-Marc Rickli. *Surrogate Warfare: Surrogate Warfare: The Transformation of War in the Twenty-First Century*. Washington, DC: Georgetown University Press, 2019.

Krishnan, Armin. "Fifth Generation Warfare, Hybrid Warfare, and Gray Zone Conflict: A Comparison." *Journal of Strategic Security* 15, no. 4 (2022): 14–31.

Kroenig, Matthew. *The Logic of American Nuclear Strategy: Why Strategic Superiority Matters*. New York, NY: Oxford University Press, 2018.

———, and Christian Trotti. "Modernization as a Promoter of International Security: The Special Role of US Nuclear Weapons." In *Nuclear Modernization in the 21st Century*, edited by Aiden Warren and Philip M. Baxter. 176-190. London: Routledge, 2020.

———, Mark J. Massa, and Christian Trotti. "Russia's Exotic Nuclear Weapons and Implications for the United States and NATO." Atlantic Council, March 6, 2020. https://www.atlanticcouncil.org/in-depth-research-reports/issue-brief/russias-exotic-nuclear-weapons-and-implications-for-the-united-states-and-nato/.

Kroenig, Matthew, Paul D. Miller, Amanda Rothschild, and Jeff Cimmino. "Scowcroft Strategy Scorecard: What grade Does Biden's

National Security Strategy Get?" *Atlantic Council* (blog), October 14, 2022. https://www.atlanticcouncil.org/content-series/scorecard/scowcroft-strategy-scorecard-what-grade-does-bidens-national-security-strategy-Get/.

Kuhn, Thomas S. *The Structure of Scientific Revolutions.* Chicago, IL: University of Chicago Press, 1996.

Kumar, Abhishek. "The Arthaśāstra: Addressing the contemporary Relevance of an Ancient Indian Treatise on Statecraft." *US Army Command and General Staff College,* 2016.

Kumar, Raksha. "How China uses the news media as a weapon in its propaganda war against the West." *Reuters Institute,* November 2, 2021. https://reutersinstitute.politics.ox.ac.uk/news/how-china-uses-news-media-weapon-its-propaganda-war-against-west.

Kuo, Rachel, and Alice Marwick. "Critical disinformation studies: History, power, and politics." *Misinformation Review, Harvard Kennedy School,* August 12, 2021.

Kupchan, Cliff. "Bipoliarity Is Back: Why It Matters." *The Washington Quarterly* 44, issue 4, (February 2, 2022): 123–139. https://doi.org/10.1080/0163660X.2021.2020457.

Kurecic, Petar. "Geoeconomic and Geopolitical Conflicts: Outcomes of the Geopolitical Economy in a Contemporary World." *World Review of Political Economy* 6, no. 4 (2015): 522–543. https://doi.org/10.13169/worlrevipoliecon.6.4.0522.

Kurlantzick, Joshua. *Beijing's Global Media Offensive: China's Uneven Campaign to Influence Asia and the World.* New York, NY: Oxford University Press, 2023.

———. "Reforming the U.S. International Military Education and Training Program." Washington, DC: Council on Foreign Relations, June 8, 2016, https://www.cfr.org/report/reforming-us-international-military-education-and-training-program.

Kuznetsova, Elizaveta. "Putin Shuts Down Russia's Free Press for Reporting Accurately on Ukraine." *Nieman Reports,* March 9, 2022. https://niemanreports.org/articles/putin-ukraine-russia-media/.

Kwok, Alan H., Emma E.H. Doyle, Julia Becker, David Johnston, and Douglas Paton. "What Is 'Social Resilience'? Perspectives of Disaster Researchers, Emergency Management Practitioners, and Policymakers in New Zealand." *International Journal of Disaster Risk Reduction* 19 (October 1, 2016): 197–211. https://doi.org/10.1016/j.ijdrr.2016.08.013.

Kwon, Jaebeom. "Taming Neighbors: Exploring China's Economic Statecraft to Change Neighboring Countries' Policies and Their Effects." *Asian Perspective* 44, no. 1, (2020): 103–138. https://muse.jhu.edu/article/748249.

LaBelle, Michael Carnegie. "Energy as a Weapon of War: Lessons from 50 Years of Energy Interdependence." *Global Policy* 14, no. 3 (June 1, 2023): 531–547. https://doi.org/10.1111/1758-5899.13235.

Lagarde, Christine. "Updating Bretton Woods: In a letter to the next generation, Christine Lagarde calls for a renewed commitment to global economic cooperation." *Finance & Development* 56, no. 1, (March 2019), Gale Academic OneFile. https://link-gale-com.proxy.library.georgetown.edu/apps/doc/A601658296/AONE?u=wash43584&sid=bookmark-AONE&xid=d3df73b9.

Lanoszka, Alexander. "Russian Hybrid Warfare and Extended Deterrence in Eastern Europe." *International Affairs* 92, no. 1, (2016): 175–195.

Lantis, Jeffrey S. "Strategic Culture and National Security Policy." *International Studies Review* 4, no. 3, (2002): 87–113.

Larson, Deborah Welch. "The Role of Belief Systems and Schemas in Foreign Policy Decision-Making." *Political Psychology* 15, no. 1 (March 1994): 17–33.

Larsen, Jeffrey A. *Comparative Strategic Cultures Curriculum: Assessing Strategic Culture as a Methodological Approach to Understanding WMD Decision-Making by States and Non-State Actors*. Washington, DC: Defense Threat Reduction Agency, October 31, 2006.

Larson, Eric V., Richard E. Darilek, Daniel Gibran, Brian Nichiporuk, Amy Richardson, Lowell H. Schwartz, and Cathryn Quantic Thurston. *Foundations of Effective Influence Operations: A Framework for*

Enhancing Army Capabilities. Santa Monica, CA: RAND, 2009. https://www.rand.org/pubs/monographs/MG654.html.

Larsson, Robert L. *Russia's Energy Policy: Security Dimensions and Russia's Reliability as an Energy Supplier.* Stockholm, FOI – Swedish Defence Research Agency: March 2006.

Lasswell, Harold Dwight. *Politics: Who Gets What, When, How.* New York, NY: Meridian Books, 1958.

Lawson, Marian Leonardo. "Does Foreign Aid Work? Efforts to Evaluate U.S. Foreign Assistance." Washington, DC: Congressional Research Service, June 23, 2016. https://sgp.fas.org/crs/row/R42827.pdf.

Le Gargasson, Clara, and James Black. "Reflecting on One Year of War: The Role of Non-Military Levers." *Center for Maritime Strategy*, February 23, 2023. https://centerformaritimestrategy.org/publications/reflecting-on-one-year-of-war-the-role-of-non-military-levers/.

Lee, John. *Chinese Political Warfare.* Washington, DC: Hudson Institute, June 23, 2022. https://www.hudson.org/foreign-policy/chinese-political-warfare-the-pla-s-information-and-influence-operations;

Legro, Jeffrey. "Military Culture and Inadvertent Escalation in World War II." *International Security* 18, no. 4 (Spring 1994): 108–142.

Leonard, Mark. "China Is Ready for a World of Disorder." *Foreign Affairs* 102, no. 4 (July/August 2023): 116–127. https://www.foreignaffairs.com/united-states/china-ready-world-disorder.

Lété, Bruno. "The Paris Call and Activating Global Cyber Norms." *German Marshall Fund Insights,* March 10, 2021. https://www.gmfus.org/news/paris-call-and-activating-global-cyber-norms.

Levinger, Matthew. "Master Narratives of Disinformation Campaigns." *Journal of International Affairs* 71, no. 1.5 (2018): 125–134. https://www.jstor.org/stable/26508126.

Lew, Jacob J., and Richard Nephew. "The Use and Misuse of Economic Statecraft: How Washington Is Abusing Its Financial Might." *Foreign Affairs* 97, no. 6 (2018): 139–149. https://www.jstor.org/stable/26797939.

Li, Shaomin. *The Rise of China, Inc.* Cambridge, UK: Cambridge University Press, 2022.

Li, Liping, and Xinzao Huang. "The Latest Research Progress of Regional Economic Resilience in China." *Proceedings of the 2022 7th International Conference on Social Sciences and Economic Development (ICSSED 2022)*, 1745–1750. https://doi.org/10.2991/aebmr.k.220405.291.

Lian, Qiao, and Wang Xiangsui. *Unrestricted Warfare: China's Master Plan to Destroy America.* Brattleboro, VT: Echo Point Books & Media, 2015.

Libiseller, Chiara. "'Hybrid Warfare' as an Academic Fashion." *Journal of Strategic Studies* 46, no. 4, (2023): 1–23.

Liebig, Michael. "India's Strategic Culture and Its Kautilyan Lineage." *The United Service Institution of India* (blog), 7, October 2020, https://www.usiofindia.org/publication-journal/indias-strategic-culture-and-its-kautilyan-lineage.html.

Linden, Sander van der, Jon Roozenbeek, and Josh Compton. "Inoculating Against Fake News About COVID-19." *Frontiers in Psychology* 11 (2020). https://www.frontiersin.org/articles/10.3389/fpsyg.2020.566790.

Linetsky, Zuri. "China Can't Catch a Break in Asian Public Opinion." *Foreign Policy*, June 28, 2023. https://foreignpolicy.com/2023/06/28/china-soft-power-asia-culture-influence-korea-singapore/.

Lippman, Daniel, Lara Seligman, Alexander Ward, and Quint Forgey. "Biden's Era of 'Strategic Competition.'" *POLITICO*, October 5, 2021. https://www.politico.com/newsletters/national-security-daily/2021/10/05/bidens-era-of-strategic-competition-494588.

Lisedunetwork. "Information: Types of information." *LIS Education Network* (blog), February 4, 2014, updated April 17, 2023. https://www.lisedunetwork.com/definition-and-types-of-information/.

Liu, Mingfu. *The China Dream: Great Power Thinking and Strategic Posture in the Post-American Era.* Beijing, CN Times Beijing Media Time United Publishing Company Limited: 2015.

Bibliography

Lock, Edward. "Refining strategic culture: return of the second generation." *Review of International Studies* 36, no. 3, (July 2010): 685–708.

Lofgren, Eric, Whitney M. McNamara, and Peter Modigliani. *Atlantic Council Commission on Defense Innovation Adoption.* Washington, DC: Atlantic Council, April 12, 2023. https://www.atlanticcouncil.org/in-depth-research-reports/report/atlantic-council-commission-on-defense-innovation-adoption-interim-report/.

Lowy Institute. "Lowy Institute Asia Power Index: 2023 Edition." https://power.lowyinstitute.org/.

Lucas, Scott, and Kaeten Mistry. "Illusions of Coherence: George F. Kennan, US Strategy and Political Warfare in the Early Cold War, 1946–1950." *Diplomatic History* 33, no. 1 (January 2009): 39–66.

Lurye, Sharon. "10 Countries That Received the Most U.S. Military Aid in 2020." *US News & World Report*, May 18, 2022. https://www.usnews.com/news/best-countries/slideshows/countries-that-received-the-most-u-s-military-aid-in-2020.

Luttwak, Edward N. *The Rise of China vs. the Logic of Strategy.* Cambridge, MA: Belknap Press, 2012.

Ma, Wenhao. "Pew Survey: Global Opinion of China Has Deteriorated Under Xi." VOA, September 30, 2022. https://www.voanews.com/a/pew-survey-global-opinion-of-china-has-deteriorated-under-xi-/6770773.html.

Maan, Ajit, and Amar Cheema, eds. *Soft Power on Hard Problems: Strategic Influence in Irregular Warfare.* Lanham, MD: Hamilton Books, 2016.

MacDonald, Justin. "Russian Abuses in Ukraine: A Lesson in Legitimacy." *The War Room, The US Army War College* (Blog), July 21, 2023. https://warroom.armywarcollege.edu/articles/legitimacy/.

Macikenaite, Vida. "China's economic statecraft: the use of economic power in an interdependent world." *Journal of Contemporary East Asia Studies* 9, no. 2, (2020): 108–126. DOI: 10.1080/24761028.2020.1848381.

Maclean, Kirsten, Michael Cuthill, and Helen Ross. "Six Attributes of Social Resilience." *Journal of Environmental Planning and Management*

57, no. 1 (January 1, 2014): 133-150. https://doi.org/10.1080/09640568.2013.763774.

Mahnken, Thomas G. "Containment: Myth and Metaphor." In *The Power of the Past*, edited by Hal Brands and Jeremy Suri, 144-156. Washington, DC: Brookings Institute Press, 2016.

———. "United States Strategic Culture." Washington, DC: Defense Threat Reduction Agency, November 13, 2006. https://irp.fas.org/agency/dod/dtra/us.pdf.

Maine, Henry Sumner. *International Law: A Series of Lectures Delivered Before the University of Cambridge, 1887.* J. Murray, 1888.

Marat, Erica. "China's Expanding Military Education Diplomacy in Central Asia." *PONARS Eurasia*, April 19, 2021. https://www.ponarseurasia.org/chinas-expanding-military-education-diplomacy-in-central-asia/.

March, James G., ed. *Influence, Leadership, Control.* New York, NY: Routledge, 2013.

Marks, Thomas A., and David H. Ucko. "Counterinsurgency as Fad: America's Rushed Engagement with Irregular Warfare." *Journal of Strategic Studies Ahead of Print* (2023): 1–27.

Markusen, Max. "A Stealth Industry: The Quiet Expansion of Chinese Private Security Companies." *CSIS*, January 12, 2022. https://www.csis.org/analysis/stealth-industry-quiet-expansion-chinese-private-security-companies.

Marrow, Alexander. "Russia's SWIFT Alternative Expanding Quickly This Year, Central Bank Says." Reuters, September 23, 2022. https://www.reuters.com/business/finance/russias-swift-alternative-expanding-quickly-this-year-says-cbank-2022-09-23/.

Martel, William C. "America's Dangerous Drift." *The Diplomat*, February 25, 2013. https://thediplomat.com/2013/02/americas-dangerous-drift/.

Martin, Timothy W. "China Relies on U.S., Allies for Hundreds of Products." *The Wall Street Journal*, August 9, 2023. https://www.wsj.com/articles/u-s-allies-could-counter-chinas-economic-coercionif-they-unite-c45d2f8d.

Masters, Jonathan, and Will Merrow. "How Much Aid Has the U.S. Sent Ukraine? Here Are Six Charts." Council on Foreign Relations, September 21, 2023. https://www.cfr.org/article/how-much-aid-has-us-sent-ukraine-here-are-six-charts.

Matloff, Maurice. "Allied Strategy in Europe, 1939–1945." In *Makers of Modern Strategy: From Machiavelli to the Nuclear Age*, edited by Peter Paret, 677–702. Princeton NJ: Princeton University Press, 1986.

Matthijs, Matthias, and Sophie Meunier. "Europe's Geoeconomic Revolution: How the EU Learned to Wield Its Real Power." *Foreign Affairs* 102, No. 5 (September/October 2023): 168–179

Mattis, Peter. "China's 'Three Warfares' in Perspective." *War on the Rocks*, January 30, 2018. https://warontherocks.com/2018/01/chinas-three-warfares-perspective/.

———. "Form and Function of the Chinese Communist Party." *Research and Analysis at Special Competitive Studies Project* (blog), September 28, 2017. https://www.linkedin.com/pulse/form-function-chinese-communist-party-peter-mattis/.

Mayer-Schönberger, Viktor, and Thomas Ramge. *Reinventing Capitalism in the Age of Big Data*. New York, NY: Basic Books, 2018.

Mazarr, Michael J. *Mastering the Gray Zone*. Carlisle, PA: U.S. Army War College, 2016.

———, Bryan Frederick, and Yvonne K. Crane. *Understanding a New Era of Strategic Competition*. Santa Monica, CA: RAND, 2022. https://www.rand.org/pubs/research_reports/RRA290-4.html.

Mazzucato, Mariana. *The Entrepreneurial State*. New York, NY: PublicAffairs, 2015.

McBridge, James, Noah Berman, and Andrew Chatzsky. "China's Massive Belt and Road Initiative." Council on Foreign Relations, February 2, 2023. https://www.cfr.org/backgrounder/chinas-massive-belt-and-road-initiative#chapter-title-0-1

McCormick, David H., Charles E. Luftig, and James M. Cunningham. "Economic Might, National Security, and the Future of American Statecraft." *Texas National Security Review* 3, no. 3 (Autumn 2020): 51–

75. https://tnsr.org/wp-content/uploads/2020/05/04_TNSR-Journal-Vol-3-Issue-3-McCormick.pdf.

McDonald, Scott D., and Michael C. Burgoyne, eds. *China's Global Influence: Perspectives and Recommendations*, Washington, DC: National Defense University Press, 2020.

McDowell, Daniel. "Financial Sanctions and Political Risk in the International Currency System." *Review of International Political Economy* 28, no. 3 (May 4, 2021): 635–661 https://doi.org/10.1080/0 9692290.2020.1736126.

McFate, Sean. "Irregular warfare with China, Russia: Ready or not, it's coming — if not already here." *The Hill*, October 11, 2020, https://thehill.com/opinion/national-security/519948-irregular-warfare-with-china-russia-ready-or-not-its-coming-if-not/.

McInnis, Kathleen J., and Clementine G. Starling. *The Case for a Comprehensive Approach 2.0* (June 2021). https://www.atlanticcouncil.org/wp-content/uploads/2021/06/NATO-Comprehensive-Approach-Report-2021.pdf.

McMann, Jason, and Jon Reid. "US Foreign Policy Tracker." *Morning Consulting Pro*, October 26, 2023. https://pro.morningconsult.com/trackers/public-opinion-us-foreign-policy.

McMaster, H.R. *Battlegrounds: The Fight to Defend the Free World*. New York, NY: Harper, 2020.

Mead, Walter Russell. *Special Providence: American Foreign Policy and How It Changed the World*. 1st Edition, New York: Routledge, 2002.

Mearsheimer, John J. "The False Promise of International Institutions." *International Security* 19, no. 3, (1994): 5–49.

———. *The Tragedy of Great Power Politics*; New York, NY: W.W. Norton & Company, 2014

Mecklin, John. "Introduction: The Evolving Threat of Hybrid War." *Bulletin of the Atomic Scientists* 73, no. 5, (2017): 298–299.

Meese, Michael J., Suzanne C. Nielsen, and Rachel M. Sondheimer. *American National Security*. Baltimore, MD: Johns Hopkins University Press, 2018.

Meierdin, Emily, and Rachel Sigman. "Understanding the Mechanisms of International Influence in an Era of Great Power Competition." *Journal of Global Security Studies* 6, no. 4 (December 1, 2021): 1–18. https://doi.org/10.1093/jogss/ogab011.

Mejias, Ulises A., and Nikolai E. Vokuev. "Disinformation and the Media: The Case of Russia and Ukraine." *Media, Culture & Society* 39, no. 7 (October 1, 2017): 1027–1042. https://doi.org/10.1177/0163443716686672.

Middaugh, James S. "Homeward Bound: World War II Ends in the Pacific." *Naval History and Heritage Command*, July 2020. https://www.history.navy.mil/browse-by-topic/wars-conflicts-and-operations/world-war-ii/1945/victory-in-pacific/homeward-bound.html.

Miron, Marina. "Russian 'Hybrid Warfare' and the Annexation of Crimea: The Modern Application of Soviet Political Warfare." *International Affairs* 99, no. 1, (2023): 397–398.

Miscamble, William. *George F. Kennan and the Making of American Foreign Policy, 1947–1950.* Princeton, NJ: Princeton University Press, 1993.

Mittelmark, Cary. "Playing Chess with the Dragon: Chinese-U.S. Competition in the Era of Irregular Warfare." *Small Wars & Insurgencies* 32, no. 2 (2021): 205–228.

Montgomery, Evan Braden. "Contested Primacy in the Western Pacific: China's Rise and the Future of U.S. Power Projection." *International Security* 38, no. 4 (2014): 115–149.

Moore, Scott M. *China's Next Act: How Sustainability and Technology are Reshaping China's Rise and the World's Future.* New York, NY: Oxford University Press, 2022.

Morgan, Pippa. "Can China's Economic Statecraft Win Soft Power in Africa? Unpacking Trade, Investment and Aid." *Journal of Chinese Political Science* 24, (2019): 387–409. https://doi.org/10.1007/s11366-018-09592-w.

Morgenstern, Emily M., and Nick M. Brown. "Foreign Assistance: An Introduction to U.S. Programs and Policy." Washington, D.C.:

Congressional Research Service, January 10, 2022. https://crsreports.congress.gov/product/pdf/R/R40213.

Morgenthau, Hans. *Politics Among Nations*. 4th Edition. New York, Ny: Alfred A. Knopf, 1967.

———. *Politics in the Twentieth Century: The Impasse of American Foreign Policy*. Chicago, IL: University of Chicago Press, 1962.

Morrell, S.C., and M.E. Kosal. "Military Deception and Strategic Culture: The Soviet Union and Russian Federation." *Journal of Information Warfare* 20, no. 3, (2021): 127–145.

Moy, Wesley, and Kacper Gradon. "COVID-19 Effects and Russian Disinformation." *Homeland Security Affairs*, no. Special COVIC-19 Issue 2020 (December 2020).

Moyer, Jonathan D., Collin J. Meisel, Austin S. Matthews, David K. Bohl, and Mathew J. Burrows. *China-US Competition: Measuring Global Influence*. Washington, DC: Atlantic Council and Frederick S. Pardee Center for International Futures, 2021. https://www.atlanticcouncil.org/wp-content/uploads/2021/06/China-US-Competition-Report-2021.pdf.

Moyer, Jonathan D., Tim Sweijs, Mathew J. Burrows, Hugo Van Manen. "IN BRIEF: Fifteen Takeaways from Our New Report Measuring US and Chinese Global Influence." Atlantic Council, June 16, 2021. https://www.atlanticcouncil.org/in-depth-research-reports/report/in-brief-15-takeaways-from-our-new-report-measuring-us-and-chinese-global-influence/.

———. "Power and Influence in a Globalized World." Atlantic Council, February 20, 2018. https://www.atlanticcouncil.org/in-depth-research-reports/report/power-and-influence-in-a-globalized-world/.

Mull, Christian, and Matthew Wallin. "Propaganda: A Tool of Strategic Influence." *American Security Project*, (September 2013).

Mumford, Andrew. "Understanding Hybrid Warfare." *Cambridge Review of International Affairs* 33, no. 6, (December 2020): 824–827.

———, and Bruno Reis, eds. *The Theory and Practice of Irregular Warfare: Warrior-Scholarship in Counter-Insurgency.* New York, NY: Routledge, 2013.

Munoz, Oscar. "Toward Trade Opening." In *Intricate Links: Democratization in Latin America and Eastern Europe*, edited by Joan M. Nelson et al., 61-104. New Brunswick: Transaction Publishing, 1994.

———, Jon Billsberry, and Veronique Ambrosini. "Resilience, Robustness and Antifragility: Towards an Appreciation of Distinct Organizational Responses to Adversity." *International Journal of Management Reviews* 24, no. 2 (April 2022): 182–183. https://onlinelibrary-wiley-com.proxy.library.georgetown.edu/doi/full/10.1111/ijmr.12289.

Murphy, Dennis, and Daniel Kuehl. "The Case for a National Information Strategy." *Military Review*, (September-October 2015): 70–83.

Murray, Williamson. "Thucydides: Theorist of War." *Naval War College Review* 66, no. 4, (2013): 30–46.

Naftali, Tim. "Political Warfare Comes Home." *Foreign Affairs*, August 18, 2023. https://www.foreignaffairs.com/united-states/trump-indictments-political-warfare-comes-home.

Nakashima, Ellen. "U.S. Cyber Command operation disrupted internet access of Russian troll factory on day of 2018 midterms." *The Washington Post,* February 27, 2019. https://www.washingtonpost.com/world/national-security/us-cyber-command-operation-disrupted-internet-access-of-russian-troll-factory-on-day-of-2018-midterms/2019/02/26/1827fc9e-36d6-11e9-af5b-b51b7ff322e9_story.html.

Nantulya, Paul. "Chinese Professional Military Education for Africa: Key Influence and Strategy." U.S. Institute of Peace, July 5, 2023, https://www.usip.org/publications/2023/07/chinese-professional-military-education-africa-key-influence-and-strategy.

Nasu, Hitoshi. "The 'Infodemic': Is International Law Ready to Combat Fake News in the Age of Information Disorder?" *The Australian Year Book of International Law Online* 39, no.1, (2021): 65–77. https://doi.org/10.1163/26660229-03901006

Nathan, Andrew J. "The Puzzle of Authoritarian Legitimacy." *Journal of Democracy* 31, no. 1, (January 2020): 158–168.

———. "The Tiananmen Papers." *Foreign Affairs*, January 1, 2001. https://www.foreignaffairs.com/articles/asia/2001-01-01/tiananmen-papers.

———, and Andrew Scobell. "How China Sees America: The Sum of Beijing's Fears." *Foreign Affairs* 91, no. 5 (October 2012): 32–47. https://www.jstor.org/stable/41720859.

National Archives. "Declaration of Independence: A Transcription." *American Founding Documents*, November 1, 2015, https://www.archives.gov/founding-docs/declaration-transcript.

———. "The Truman Doctrine," *Milestone Documents*. https://www.archives.gov/milestone-documents/truman-doctrine.

National Guard. "State Partnership Program." https://www.nationalguard.mil/leadership/joint-staff/j-5/international-affairs-division/state-partnership-program/.

National Intelligence Council. "Global Trends: The Paradox of Progress" (January 2017). https://www.dni.gov/files/documents/nic/GT-Full-Report.pdf.

"National Security Act of 1947," (Public Law 235 of July 26, 1947; 61 STAT. 496). IC Legal Reference Book, Office of the Director of National Intelligence. https://www.dni.gov/index.php/ic-legal-reference-book/national-security-act-of-1947.

NATO. "The North Atlantic Treaty," April 4, 1949. https://www.nato.int/cps/en/natolive/official_texts_17120.htm#:~:text=Article%202,of%20stability%20and%20well%2Dbeing.

———. "Resilience, Civil Preparedness and Article 3." August 2, 2023. https://www.nato.int/cps/en/natohq/topics_132722.htm.

Nelson, Joan M. *Intricate Links: Democratization and Market Reforms in Latin America and Eastern Europe*. London: Routledge, 1994.

Nelson, Rebecca M. et al., *Russia's War on Ukraine: Financial and Trade Sanctions*. Congressional Research Service Report IF12062, updated February 22, 2023. https://crsreports.congress.gov/product/pdf/IF/IF12062.

Nelson, Rebecca M., and Martin A. Weiss, "The U.S. Dollar as the World's Dominant Reserve Currency" (Washington, D.C.: Congressional Research Service, September 15, 2022), https://crsreports.congress.gov/product/pdf/IF/IF11707.

Nemeth, William J. *Future War and Chechnya: A Case for Hybrid Warfare*. Monterey, CA: Naval Postgraduate School, 2002.

Nexon, Daniel H. "Against Great Power Competition." *Foreign Affairs*, February 15, 2021. https://www.foreignaffairs.com/articles/united-states/2021-02-15/against-great-power-competition.

Ngo, Tri Minh, and Koh Swee Lean Collin. "Lessons from the Battle of the Paracel Islands." *The Diplomat*, January 23, 2014. https://thediplomat.com/2014/01/lessons-from-the-battle-of-the-paracel-islands/.

Ngoun, Kimly. "Adaptive Authoritarian Resilience: Cambodian Strongman's Quest for Legitimacy." *Journal of Contemporary Asia* 52, no. 1 (January 1, 2022): 23–44. https://doi.org/10.1080/00472336.2020.1832241.

Nguyen, Xuan Quynh et al., "China Is Winning the Silent War to Dominate the South China Sea," *Bloomberg*, July 10, 2019. https://www.bloomberg.com/graphics/2019-south-china-sea-silent-war/

Ni, Vincent. "China Funnels Its Overseas Aid Money into Political Leaders' Home Provinces." *The Guardian*, May 29, 2022. https://www.theguardian.com/world/2022/may/29/china-funnels-overseas-aid-money-political-leaders-home-provinces.

Nilsson, Mikael. "American Propaganda, Swedish Labor, and the Swedish Press in the Cold War: The United States Information Agency (USIA) and Co-Production of U.S. Hegemony in Sweden during the 1950s and 1960s." *The International History Review* 34, vol. 2, (June 1, 2012): 315–345.

Norris, Fran H., Susan P. Stevens, Betty Pfefferbaum, Karen F. Wyche, and Rose L. Pfefferbaum. "Community Resilience as a Metaphor, Theory, Set of Capacities, and Strategy for Disaster Readiness." *American Journal of Community Psychology* 41, no. 1–2 (2008): 127–150. https://doi.org/10.1007/s10464-007-9156-6.

Nye, Joseph S., Jr. *The Future of Power*. New York, NY: PublicAffairs, 2011.

———. "Is America Reverting to Isolationism?" *Project Syndicate*, September 4, 2023. https://www.project-syndicate.org/commentary/us-republicans-dangerous-isolationism-by-joseph-s-nye-2023-09.

———. "Public Diplomacy and Soft Power." *The ANNALS of the American Academy of Political and Social Science* 616, no. 1 (March 2008): 94–109.

———. "Soft Power and American Foreign Policy." *Political Science Quarterly* 119, no. 2 (2004): 255–270

———, and William A. Owens. "America's Information Edge." *Foreign Affairs*, February 1, 1996. https://www.foreignaffairs.com/articles/united-states/1996-03-01/americas-information-edge

O'Brien, Phillips. "The War That Defied Expectations: What Ukraine Revealed About Military Power." *Foreign Affairs*, July 27, 2023. https://www.foreignaffairs.com/ukraine/war-defied-expectations.

O'Hanlon, Michael. "China's Shrinking Population and Constraints on Its Future Power." *The Brookings Institution*, April 24, 2023. https://www.brookings.edu/articles/chinas-shrinking-population-and-constraints-on-its-future-power/.

Obrist, Brigit, Constanze Pfeiffer, and Bob Henley. "Multi-Layered Social Resilience: A New Approach in Mitigation Research." *Progress in Development Studies* 10 (September 1, 2010): 283–293. https://doi.org/10.1177/146499340901000402.

OECD. *Building Trust and Reinforcing Democracy: Preparing the Ground for Government Action*, OECD Public Governance Reviews. Paris: OECD Publishing, 2022. https://doi.org/10.1787/76972a4a-en.

Office of Inspector General. *DHS Needs a Unified Strategy to Counter Disinformation Campaigns.* https://www.oig.dhs.gov/sites/default/files/assets/2022-08/OIG-22-58-Aug22.pdf.

Office of the Under Secretary of Defense (Comptroller)/Chief Financial Officer. April 2022, Defense Budget Overview, United States Department of Defense Fiscal Year 2023 Budget Request.

Office of the Under Secretary of Defense (Comptroller)/Chief Financial Officer, *Pacific Deterrence Initiative: Department of Defense Budget Fiscal Year (FY) 2023.* Arlington: Department of Defense, 2022. https://

comptroller.defense.gov/Portals/45/Documents/defbudget/FY2023/ FY2023_Pacific_Deterrence_Initiative.pdf.

Office of the Director of National Intelligence. *Assessing Russian Activities and Intentions in Recent US Elections.* https://www.intelligence.senate. gov/sites/default/files/documents/ICA_2017_01.pdf.

Oldfield, Jackson. "The Challenges of Asset Freezing Sanctions as an Anti-Corruption Tool." *Transparency International Anti-Corruption Helpdesk*, 2022. https://knowledgehub.transparency.org/helpdesk/the-challenges-of-asset-freezing-sanctions-as-an-anti-corruption-tool.

Olszewski, Thomas, Irina Liu, and Allison Cunningham. "Survey of Federal Community Resilience Programs and Available Resilience Planning Tools." National Institute of Standards and Technology, February 18, 2021. https://doi.org/10.6028/NIST.GCR.21-027.

Orden, Hedvig, and James Pamment. "What Is So Foreign About Foreign Influence Operations?" *Lines in the Sand Series.* Carnegie Endowment for International Peace, January 2021.

Orsini, Ryan J. "Economic Statecraft and US-Russian Policy." *The US Army War College Quarterly: Parameters* 52, no. 2 (May 18, 2022): 35–54. https://doi.org/10.55540/0031-1723.3151.

Ostflaten, Amund. "Russian Strategic Culture after the Cold War The Primacy of Conventional Force." *Journal of Military & Strategic Studies* 20, no. 2, (October 2019): 110–132.

Ota, Fumio. "Sun Tzu in Contemporary Chinese Strategy." *Joint Forces Quarterly* 73, no. 2, 2014. https://ndupress.ndu.edu/JFQ/ Joint-Force-Quarterly-73/Article/577507/sun-tzu-in-contemporary-chinese-strategy/.

Overfield, Cornell. "Biden's 'Strategic Competition' Is a Step Back." *Foreign Policy*, October 13, 2021. https://foreignpolicy.com/2021/10/1 3/biden-strategic-competition-national-defense-strategy/.

Owen, Geoffrey. *Lessons from the US: Innovation Policy*, London: Policy Exchange, 2017.

Owens, Mackubin Thomas. "American Strategic Culture and Civil-Military Relations: The Case of JCS Reform." *Naval War College Review* 39, no. 2 (April 1986): 43–59.

Painter, David. "Energy and the End of the Evil Empire." In *The Reagan Moment: America and the World in the 1980s*, edited by Jonathan R. Hunt and Simon Miles, 43-63. Ithaca NY: Cornell University Press, 2021.

Paris Call For Trust and Security in Cyberspace. "Home." https://pariscall.international/en/.

Parker, Ned, and Peter Eisler, "Political violence in polarized US at its worst since 1970s." Reuters, August 9, 2023. https://www.reuters.com/investigates/special-report/usa-politics-violence/.

Pascu, I.M., and N.-S. Vintila. "Strengthening the Resilience of Political Institutions and Processes: A Framework of Analysis." Connections QJ 19, no. 3 (2020): 55–66, https://doi.org/10.11610/Connections.19.3.04.

Paterson, Pat. "The Truth About Tonkin." *Naval History Magazine* 22, no. 1, (February 2008). https://www.usni.org/magazines/naval-history-magazine/2008/february/truth-about-tonkin.

Paul, Christopher M., Michale Schwille, Michael Vasseur, Elizabeth M. Bartels, and Ryan Bauer. *The Role of Information in U.S. Concepts for Strategic Competition*. Santa Monica, CA: RAND, 2022. https://www.rand.org/pubs/research_reports/RRA1256-1.html.

Paul, Christopher, and Miriam Matthews. "The Russian 'Firehose of Falsehood' Propaganda Model." Santa Monica, CA: RAND Corporation, July 11, 2016. https://www.rand.org/pubs/perspectives/PE198.html.

Paul, T.V. "Indian Soft Power in a Globalizing World." *Current History* 113, no. 762, (2014): 157–162.

Pavel, Barry, and Christian Trotti. "New Tech Will Erode Nuclear Deterrence. The US Must Adapt." *Defense One* (blog), November 4, 2021. https://www.defenseone.com/ideas/2021/11/new-tech-will-erode-nuclear-deterrence-us-must-adapt/186634/.

Pavel, Barry, and Daniel Egel. "A Civilian U.S. 'Joint Chiefs' for Economic Competition with China?" *RAND Blog* (blog), April 24, 2023. https://

www.rand.org/blog/2023/04/a-civilian-us-joint-chiefs-for-economic-competition.html.

Perrigo, Billy. "Why a Blank Sheet of Paper Became a Protest Symbol in China." *Time*, December 1, 2022. https://time.com/6238050/china-protests-censorship-urumqi-a4/.

Perry, Walter L., Stuart Johnson, Stephanie Pezard, Gillian S. Oak, David Stebbins, and Chaoling Feng, *Defense Institution Building: An Assessment*. Santa Monica, CA: RAND Corporation, 2016. https://www.rand.org/pubs/research_reports/RR1176.html.

Peters, Robert. "America's Current Nuclear Arsenal Was Built for a More Benign World." *Heritage Foundation*, September 14, 2023. https://www.heritage.org/defense/commentary/americas-current-nuclear-arsenal-was-built-more-benign-world.

Peterson, Dahlia. "Designing Alternatives to China's Repressive Surveillance State." Washington, DC: CSET, October 2020. https://cset.georgetown.edu/publication/designing-alternatives-to-chinas-repressive-surveillance-state/.

Peterson, Rachelle. "China's Confucius Institutes Might Be Closing, But They Succeeded." *National Association of Scholars*, March 31, 2021. https://www.nas.org/blogs/article/chinas-confucius-institutes-might-be-closing-but-they-succeeded.

Petit, Brian. "Can Ukrainian Resistance Foil a Russian Victory?" *War on the Rocks*, February 18, 2022. https://warontherocks.com/2022/02/can-ukrainian-resistance-foil-a-russian-victory/.

Pfefferbaum, Rose L., Barbara R Neas, Betty Pfefferbaum, Fran H Norris, and Richard L Van Horn. "The Communities Advancing Resilience Toolkit (CART): Development of a Survey Instrument to Assess Community Resilience." *International Journal of Emergency Mental Health* 15, no. 1 (2013): 15–29. http://europepmc.org/abstract/MED/24187884.

Pfefferbaum, Rose L., Betty Pfefferbaum, Pascal Nitiéma, J. Brian Houston, and Richard L. Van Horn. "Assessing Community Resilience: An Application of the Expanded CART Survey Instrument With Affiliated Volunteer Responders." *American Behavioral Scientist* 59, no. 2

(February 1, 2015): 181–199. https://doi.org/10.1177/0002764214550 295.

Phillips, Matt. "China's Currency is Weakening as the Trade War Drags On." *New York Times,* August 27, 2019. https://www.nytimes.com/2 019/08/27/business/china-yuan.html.

Polyakova, Alina, and Daniel Fried. "Democratic Defense Against Disinformation 2.0." Atlantic Council Eurasia Center, June 2019.

Polyakova, Alina, Flemming Splidsboel Hansen, Robert Van Der Noordaa, Øystein Bogen, and Henrik Sundbom. *The Kremlin's Trojan Horses 3.0: Russian Influence in Denmark, The Netherlands, Norway, and Sweden.* Washington, DC: Atlantic Council, Eurasia Center, 2018. https://www.atlanticcouncil.org/wp-content/uploads/2021/02/The-Kremlins-Trojan-Horses-3.pdf.

Polyakova, Alina, Marlene Laruelle, Stefan Meister, and Neil Barnett. *The Kremlin's Trojan Horses: Russian Influence in France, Germany, and the United Kingdom.* 3rd Edition. Washington, DC: Atlantic Council, Eurasia Center, 2016. https://www.atlanticcouncil.org/wp-content/uploads/2016/11/The_Kremlins_Trojan_Horses_web_0228_third_edition.pdf.

Porche III, Isaac R. et al., "Redefining Information Warfare Boundaries for an Army in A Wireless World." Santa Monica, CA: RAND Corporation, 2013. https://www.rand.org/content/dam/rand/pubs/monographs/MG1100/MG1113/RAND_MG1113.pdf.

Posen, Barry R. "Command of the Commons: The Military Foundation of U.S. Hegemony." *International Security* 28, no. 1 (2003): 5–46.

Poushter, Jacob, Moira Fagan, Sneha Gubbala and Jordan Lippert. "Americans Hold Positive Feelings Toward NATO, Ukraine and See Russia as an Enemy." *Pew Research Center,* May 10, 2023. https://www.pewresearch.org/global/2023/05/10/americans-hold-positive-feelings-toward-nato-and-ukraine-see-russia-as-an-enemy/.

Preston, Andrew, ed. *Rethinking American Grand Strategy,* New York, NY: Oxford University Press, 2021.

Prier, Jarred. "Commanding the Trend: Social Media as Information Warfare." *Strategic Studies Quarterly* 11, no. 4 (2017): 50–85.

Prior, Tim. "Resilience: The 'Fifth Wave' in the Evolution of Deterrence." Center for Security Studies (CSS), ETH Zürich, 2018. https://doi.org/10.3929/ethz-b-000317733.

Pronk, Danny. "The Return of Political Warfare." *Strategic Monitor* (blog), 2019. https://www.clingendael.org/pub/2018/strategic-monitor-2018-2019/the-return-of-political-warfare/.

Pye, Lucian W., and Mary W. Pye. *Asian Power and Politics: The Cultural Dimensions of Authority.* Cambridge, MA: Belknap Press, 1985.

Querze, Alana Renee. "Can You Spend Your Way to Victory? The Case of Statebuilding in Afghanistan." PhD diss., University of Kansas, 2011.

Quinlan, Allyson E., Marta Berbes-Blazquez, L. Jamila Haider, and Garry D. Peterson. "Measuring and Assessing Resilience: Broadening Understanding through Multiple Disciplinary Perspectives." *Journal of Applied Ecology* 53, no. 3 (June 2016): 677–687. https://www.jstor.org/stable/43869698.

Rampe, William. "What Is Russia's Wagner Group Doing in Africa?" Council on Foreign Relations, May 23, 2023. https://www.cfr.org/in-brief/what-russias-wagner-group-doing-africa#:~:text=The%20Wagner%20Group%20has%20established,resource%20concessions%20and%20diplomatic%20support.

Raven, Bertram H., Joseph Schwarzwald, and Meni Koslowsky. "Conceptualizing and Measuring a Power/Interaction Model of Interpersonal Influence." *Journal of Applied Social Psychology* 28, no. 4 (February 1, 1998): 307–332. https://doi.org/10.1111/j.1559-1816.1998.tb01708.x.

Reagan, Ronald. "Evil Empire Speech." Speech, National Association of Evangelicals, Orlando, Florida, March 8, 1983. https://voicesofdemocracy.umd.edu/reagan-evil-empire-speech-text/.

———. "The SDI Speech." Speech, White House, Washington, DC, March 23, 1983), https://www.atomicarchive.com/resources/documents/missile-defense/sdi-speech.html.

Record, Jeffrey. "Back to the Powell-Weinberger Doctrine?" *Strategic Studies Quarterly* 1, no. 1 (Fall 2007): 79–95.

Reich, Simon, and Richard Ned Lebow, *Good-Bye Hegemony!: Power and Influence in the Global System.* Princeton, NJ: Princeton University Press, 2014.

Reik, Theodor, *Curiosities of Self.* Farrar, Straus & Giroux, 1965.

Reisher, Jonathan "The Effect of Disinformation on Democracy: The Impact of Hungary's Democratic Decline." *CES Working Papers* 14, no. 1 (2022): 42–68.

Reiter, Dan, and Allan C. Stam, *Democracies at War.* Princeton, NJ: Princeton University Press, 2002.

Reus-Smit, Christian. "Power, Legitimacy, and Order." *The Chinese Journal of International Politics* 7, no. 3 (Autumn 2014): 341–359.

Reuter, Tim. "An Isolationist United States? If Only That Were True." Forbes, October 10, 2013. https://www.forbes.com/sites/timreuter/2013/10/10/an-isolationist-united-states-if-only-that-were-true/.

Rid, Thomas. *Active Measures: The Secret History of Disinformation and Political Warfare.* New York, NY: Farrar, Straus and Giroux, 2020.

Riess, Helen. "Institutional Resilience: The Foundation for Individual Resilience, Especially During COVID-19." *Global Advances in Health and Medicine* 10 (January 1, 2021): https://doi.org/10.1177/21649561211006728.

Ronfeldt, David, and John Arquilla. *Whose Story Wins: Rise of the Noosphere, Noopolitik, and Information-Age Statecraft.* Santa Monica, CA: RAND Corporation, 2020. https://www.rand.org/pubs/perspectives/PEA237-1.html.

Rosenbach, Erich, and Katherine Mansted. "The Geopolitics of Information." Belfer Center for Science and International Affairs, Harvard Kennedy School, May 28, 2019.

Rubin, Victoria L. "Disinformation and misinformation triangle: A conceptual model for "fake news" epidemic, causal factors and interventions." *Journal of Documentation* 75, no. 5, (2019): 1013–1034. https://doi.org/10.1108/JD-12-2018-0209.

Rumer, Eugene, and Richard Sokolsky. "Etched in Stone: Russian Strategic Culture and the Future of Transatlantic Security."

Carnegie Endowment for International Peace, September 8, 2020. https://carnegieendowment.org/2020/09/08/etched-in-stone-russian-strategic-culture-and-future-of-transatlantic-security-pub-82657.

"Russia fights back in information war with jail warning." Reuters, March 4, 2022. https://www.reuters.com/world/europe/russia-introduce-jail-terms-spreading-fake-information-about-army-2022-03-04/

"Russia's Global Arms Exports Suffer As War Takes Toll; Ukraine's Imports Surge." Radio Free Europe/Radio Liberty, March 13, 2023. https://www.rferl.org/a/global-arms-sales-sipri-russia-ukraine/3231 4407.html.

"Russian Patriarch Kirill Says Dying In Ukraine 'Washes Away All Sins.'" Radio Free Europe/Radio Liberty, September 26, 2022. https://www.rferl.org/a/russia-patriarch-kirill-dying-ukraine-sins/32052380.html.

Ryan, Maria. "'Full spectrum dominance': Donald Rumsfeld, the Department of Defense, and US Irregular Warfare Strategy, 2001–2008." *Small Wars & Insurgencies* 25, no. 1, (2014): 41–68.

Samet, Elizabeth D., and Suzanne Toren. *Looking for the Good War: American Amnesia and the Violent Pursuit of Happiness.* Tantor Audio, 2021. (Audiobook).

Sanders, Gregory, and Asya Akca. "The Great Unwinding: The U.S.-Turkey Arms Sales Dispute." *CSIS*, March 17, 2020. https://www.csis.org/analysis/great-unwinding-us-turkey-arms-sales-dispute.

Saran, Shyam. *How India Sees the World: Kautilya to the 21st Century.* New Dehli: Juggernut, 2018.

Saressalo, T., and A. M. Huhtinen. "Information Influence Operations: Application of National Instruments of Power." *Journal of Information Warfare* 21, no. 4 (2022): 41–66.

Sargent, John F. "US Research and Development Funding and Performance: Fact Sheet." *Congressional Research Service*, September 13, 2022. https://crsreports.congress.gov/product/pdf/R/R44307.

Sarkar, Rumu. "Trends in Global Finance: The New Development (BRICS) Bank." *Loyola University Chicago International Law Review* 13, no. 2 (2016): 89–104.

Sandeman, Hugh, and Jonny Hall. "NATO's Resilience: The First and Last Line of Defence." London School of Economics and Political Science, June 2022. https://www.lse.ac.uk/ideas/publications/old-updates/nato-resilience.

Saul, John R. "The New Era of Irregular Warfare." *Queen's Quarterly* 111, no. 3, (2004): 428–431.

Saunders, Phillip C., and Jiunwei Shyy. "China's Military Diplomacy." In *China's Global Influence: Perspectives and Recommendations*, edited by Scott McDonald and Michael Burgoyne, 207–227. https://dkiapcss.edu/wp-content/uploads/2019/10/13-Chinas-Military-Diplomacy-Saunders-Shyy-rev.pdf.

Schelling, Thomas. *Arms and Influence*. Revised Edition. New Haven, CT: Yale University Press, 2008.

Scheipers, Sibylle, ed. *Heroism and the Changing Character of War*. New York: Palgrave MacMillan, 2014.

———. "Counterinsurgency or Irregular Warfare? Historiography and the Study of "Small Wars.'" *Small Wars & Insurgencies* 25, no. 5–6, (2014): 879–899.

Schoen, Fletcher, and Christopher J. Lamb. *Deception, Disinformation, and Strategic Communications: How One Interagency Group Made a Major Difference*. Strategic Perspectives no.11, Institute for National Strategic Studies, National Defense University, June 2012.

Scholivin, Soren, and Mikael Wigell. "Geoeconomic Power Politics: An Introduction." In *Geoeconomics and Power Politics in the 21st Century*, edited by Mikael Wigell et al., 1-25. London: Routledge, 2020.

Schrijvers, Peter. "War Against Evil: The Second World War." In *Heroism and the Changing Character of War*, edited by Sibylle Scheipers, 76-92. New York: Palgrave MacMillan, 2014.

Scobell, Andrew, Edmund J. Burke, Cortez A. Cooper III, Sale Lilly, Chad J. R. Ohlandt, Eric Warner, and J.D. Williams. *China's Grand Strategy: Trends, Trajectories, and Long-Term Competition*. Santa Monica, CA: RAND Corporation, 2020. https://www.rand.org/pubs/research_reports/RR2798.html.

Seldin, Jeff. "China's Speed in Selling Arms Prompts US Partners to Buy From Beijing, Say Officials." VOA, March 16, 2023. https://www.voanews.com/a/china-s-speed-in-selling-arms-prompt-us-partners-to-buy-from-beijing-say-officials/7009288.html.

Shalal, Andrea. "U.S. Wants to End Dependence on China Rare Earths, Yellen Says." Reuters, July 18, 2022. https://www.reuters.com/markets/us/us-wants-end-dependence-china-rare-earths-yellen-2022-07-18/.

Shambaugh, David, ed. *China and the World*, New York, NY: Oxford University Press, 2020.

Shane, Scott, Cade Metz, and Daisuke Wakabayashi. "How a Pentagon Contract Became an Identity Crisis for Google." *The New York Times*, May 30, 2018. https://www.nytimes.com/2018/05/30/technology/google-project-maven-pentagon.html.

Shane III, Leo, and Joe Gould. "Congress passes defense policy bill with budget boost, military justice reforms." *Defense News,* December 15, 2021. https://www.defensenews.com/news/pentagon-congress/2021/12/15/congress-passes-defense-policy-bill-with-budget-boost-military-justice-reforms/.

Shapiro, Jacob N., and Alicia Wanless, "Why Are Authoritarians Framing International Approaches to Disinformation?" *Lawfare,* December 28, 2021. https://www.lawfaremedia.org/article/why-are-authoritarians-framing-international-approaches-disinformation.

Shaw, Ryan. "In Defense of Competition." *Real Clear Defense*, November 11, 2021. https://www.realcleardefense.com/articles/2021/11/11/in_defense_of_competition_803143.html.

Sheehan, Michael A. "Army Doctrine and Irregular Warfare." Fort Leavenworth, KS: U.S. Army Command and General Staff College, 1992.

Sherman, Justin. "Untangling the Russian Web: Spies, Proxies, and Spectrums of Russian Cyber Behavior." Atlantic Council Issue Brief, September 19, 2022.

Shirk, Susan L. *Overreach: How China Derailed Its Peaceful Rise.* New York, NY: Oxford University Press, 2022.

Siegle, Joseph. "Russia and Africa: Expanding Influence and Instability." In *Russia's Global Reach: A Security and Statecraft Assessment,* edited by Graeme P. Herd, George C. Marshall European Center for Security Studies. https://www.marshallcenter.org/en/publications/marshall-center-books/russias-global-reach-security-and-statecraft-assessment/chapter-10-russia-and-africa-expanding-influence-and.

Šimonović, Ivan. "The Responsibility to Protect." UN Chronicle 4, no. 53, (December 2016).

Singer, Peter W. "Winning the War of Words: Information Warfare in Afghanistan." Brookings, October 23, 2001. https://www.brookings.edu/articles/winning-the-war-of-words-information-warfare-in-afghanistan/.

Singh, Prashant Kumar. "Rereading Mao's Military Thinking." *Strategic Analysis* 37, no. 5 (2013): 558–580.

Siriprapu, Anshu, and Noah Berman. "Is Industrial Policy Making a Comeback?" Council on Foreign Relations, November 18, 2022. https://www.cfr.org/backgrounder/industrial-policy-making-comeback.

Sisson, Melanie W., James A. Siebens, and Barry M. Blechman. *Military Coercion and US Foreign Policy: The Use of Force Short of War.* London: Routledge, 2020.

Sitaraman, Ganesh. "A Grand Strategy of Resilience: American Power in the Age of Fragility." *Foreign Affairs* 99, no. 5 (September/October 2020): 165–174. https://www.foreignaffairs.com/articles/united-states/2020-08-11/grand-strategy-resilience.

Sliwinski, Alicia. "Resilience." In *Humanitarianism: Key Words*, edited by Antonio De Lauri, 178-180. Leiden: Brill, 2020.

Skak, Mette. "Russian Strategic Culture: The Role of Today's Chekisty." *Contemporary Politics* 22, no. 3 (September 2016): 324–341.

Smeets, Martin. "Can Sanctions Be Effective? WTO Staff Working Paper ERSD-2018-03." World Trade Organization, March 15, 2018. https://www.wto.org/english/res_e/reser_e/ersd201803_e.pdf.

Smith, Bruce W., Jeanne Dalen, Kathryn Wiggins, Erin Tooley, Paulette Christopher, and Jennifer Bernard. "The Brief Resilience Scale: Assessing the Ability to Bounce Back." *International Journal of*

Behavioral Medicine 15, no. 3 (September 1, 2008): 194–200. https://doi.org/10.1080/10705500802222972.

Snyder, Jack L. *The Soviet Strategic Culture*. Santa Monica, CA: RAND, 1977. https://www.rand.org/pubs/reports/R2154.html.

Society of Professional Journalists. "Home," https://www.spj.org/.

"Somalia says Russia grants relief on debt worth $684 million." Reuters, July 27, 2023. https://www.reuters.com/world/africa/somalia-says-russia-grants-relief-debt-worth-684-million-2023-07-27/.

Spaulding, Suzanne, Emily Harding, Hadeil Ali, Joseph Majkut, Harshana Ghoorhoo, Morgan Higman, Devi Nair, Naz Subah, and Sophia Barkoff, "Innovation for Resilience." CSIS, March 23, 2023. https://www.csis.org/analysis/innovation-resilience.

Standish, Reid. "Central Asia Caught Between 'Two Fires' As It Branches Out From Russia." Radio Free Europe, December 6, 2022. https://www.rferl.org/a/central-asia-russia-china/32164503.html.

Starling, Clementine G., Tyson K. Wetzel, and Christian Trotti, *Seizing the Advantage: A Vision for the Next US National Defense Strategy*. Washington, DC: Atlantic Council, Scowcroft Center for Strategy and Security, December 2021.

Steil, Benn, and Robert E. Litan. *Financial Statecraft: The Role of Financial Markets in American Foreign Policy*. New Haven, CT: Yale University Press, 2006.

Stewart, Edward C., and Milton J. Bennett. *American Cultural Patterns: A Cross-Cultural Perspective*. 2nd Edition. Nicholas Brealey, 2011.

Stewart, Geoffrey T., Ramesh Kolluru, and Mark Smith. "Leveraging Public-private Partnerships to Improve Community Resilience in Times of Disaster." *International Journal of Physical Distribution & Logistics Management* 39, no. 5 (January 1, 2009): 343–364. https://doi.org/10.1108/09600030910973724.

Stewart, Phil, and Idrees Ali. "US Asks What's Next for Wagner Group in Middle East, Africa, After Mutiny Attempt." Reuters, June 28, 2023. https://www.reuters.com/world/us-asks-whats-next-russias-wagner-middle-east-africa-2023-06-27/.

Stewart, Richard, ed. *American Military History: The United States Army in a Global Era, 1917–2008.* Volume II. Washington, DC: Center of Military History, 2010.

Stoicescu, Kalev, Mykola Nazarov, Keir Giles, and Matthew D. Johnson. "How Russia Went to War: The Kremlin's Preparations for Its Aggression Against Ukraine." *International Centre for Defence and Security,* April 25, 2023. https://icds.ee/en/how-russia-went-to-war-the-kremlins-preparations-for-its-aggression-against-ukraine/.

Strachan, Hew. *The Direction of War: Contemporary Strategy in Historical Perspective.* Cambridge, UK: Cambridge University Press, 2013.

Strachan, Hew, and Andreas Rothe-Herberg. *Clausewitz in the Twenty-First Century.* Oxford, UK: Oxford University Press, 2007.

Strathern, Marilyn. "'Improving ratings': audit in the British University system." *European Review* 5, no. 3 (1997): 305–321.

Stuster, J. Dana. "The Strategic Logic of Russia's Bounty Policy." *Lawfare* (blog), July 10, 2020. https://www.lawfaremedia.org/article/strategic-logic-russias-bounty-policy.

Suchman, Mark C. "Managing Legitimacy: Strategic and Institutional Approaches." *The Academy of Management Review* 20, no. 3 (1995): 571–610.

Suchkov, Maxim A. "Whose Hybrid Warfare? How 'The Hybrid Warfare' Concept Shapes Russian Discourse, Military, and Political Practice." *Small Wars & Insurgencies* 32, no. 2, (April 3, 2021): 415–440.

Suits, Devon L. "Building Relationships, Interoperability through Exchange Program." U.S. Army, October 8, 2019. https://www.army.mil/article/224669/building_relationships_interoperability_through_exchange_program

Sullivan, Jake. "Remarks by National Security Advisor Jake Sullivan at the Special Competitive Studies Project Global Emerging Technologies Summit." White House, September 16, 2022. https://www.whitehouse.gov/briefing-room/speeches-remarks/2022/09/16/remarks-by-national-security-advisor-jake-sullivan-at-the-special-competitive-studies-project-global-emerging-technologies-summit/.

Sun Tzu. *The Art of War.* Translated by Michael Nylan. New York: W.W. Norton & Company, 2020.

Szalwinski, Alison, and Michael Wilson. *Understanding Strategic Cultures in the Asia-Pacific.* Washington, DC: The National Bureau of Asian Research, 2016.

Taddeo, Mariarosaria. "Just Information Warfare." *Topoi* 35, no. 1 (April 1, 2016): 213–224. https://doi.org/10.1007/s11245-014-9245-8.

Talent, Jim, and Pete Hegseth. "America's Strategic Drift." *National Review*, October 6, 2014. https://www.nationalreview.com/2014/10/americas-strategic-drift-jim-talent-pete-hegseth/.

Tarbert, Heath P. "Modernizing CFIUS." *George Washington Law Review* 88, no. 6 (November 2020): 1477–1524.

Taylor, Margaret L. "Combating Disinformation and Foreign Interference in Democracies: Lessons from Europe." Brookings, July 31, 2019. https://www.brookings.edu/articles/combating-disinformation-and-foreign-interference-in-democracies-lessons-from-europe/.

Tellis, Ashley J., Janice Bially, Christopher Layne, and Melissa McPherson. "Measuring National Power in the Postindustrial Age." RAND Corporation, January 1, 2000, 1–33. https://www.rand.org/pubs/monograph_reports/MR1110.html.

Theohary, Catherine A. "Information Warfare: Issues for Congress." Washington, DC: Congressional Research Service, March 5, 2018.

Thomas, Timothy L. *The Chinese Way of War.* McLean, VA: MITRE CORP, June 1, 2020. https://apps.dtic.mil/sti/citations/AD1114504.

Tian, Nan, Diego Lopes da Silva, Xiao Liang, Lorenzo Scarazzato, Lucie Béraud-Sudreau, and Ana Carolina de Oliveira Assis. "Trends in World Military Expenditure, 2022." *SIPRI*, April 2023, https://www.sipri.org/sites/default/files/2023-04/2304_fs_milex_2022.pdf.

"Timeline: Broadcom-Qualcomm saga comes to an abrupt end." Reuters, March 14, 2018, https://www.reuters.com/article/us-qualcomm-m-a-broadcom-timeline/timeline-broadcom-qualcomm-saga-comes-to-an-abrupt-end-idUSKCN1GQ22N.

Tocqueville, Alexis de. *Democracy in America and Two Essays on America*. Edited by Isaac Kramnick, translated by Gerald Bevan. New York: Penguin, 2003.

Trade Partnership Worldwide. *Trade and American Jobs: The Impact of Trade on U.S. and State-Level Employment: 2019 Update.* https://tradepartnership.com/reports/trade-and-american-jobs-the-impact-of-trade-on-u-s-and-state-level-employment-update-2019/.

Triggs, AJ, and McKibbin, WJ. "Global implications of a US-led currency war." *World Econ.* 44 (2021): 1484–1508. https://doi.org/10.1111/twec.13112.

Trotti, Christian. "The Return of Great-Power Hubris." *Georgetown Security Studies Review* (blog), December 12, 2022. https://georgetownsecuritystudiesreview.org/2022/12/12/the-return-of-great-power-hubris/.

———. "To Centralize or Not to Centralize? For Defense Budgeting, That Is the (Unwritten) Question.," *Georgetown Security Studies Review* (blog), May 8, 2023., https://georgetownsecuritystudiesreview.org/2023/05/08/to-centralize-or-not-to-centralize-for-defense-budgeting-that-is-the-unwritten-question/.

Tucker, Patrick. "As Irregular Warfare Comes to a Crossroads, Congress Chips in." *Defense One,* December 17, 2023. https://www.defenseone.com/policy/2023/12/irregular-warfare-comes-crossroads-congress-chips/392821/.

"Turkey rejects U.S. troop proposal." CNN, March 1, 2003. https://www.cnn.com/2003/WORLD/meast/03/01/sprj.irq.main/.

"Turkey's Russian Air Defense System and the U.S. Response." Reuters, October 1, 2021. https://www.reuters.com/world/europe/turkeys-russian-air-defence-systems-us-response-2021-10-01/.

"U.N. votes to end US embargo on Cuba; US and Israel oppose." Reuters, November 2, 2023. https://www.reuters.com/world/americas/un-votes-end-us-embargo-cuba-us-israel-oppose-2023-11-02/.

U.S. Agency for Global Media. "Mission." https://www.usagm.gov/who-we-are/mission/.

———. "Truth over Disinformation: Supporting Freedom and Democracy" (USAGM Strategic Plan 2022–2026). https://www.usagm.gov/wp-content/uploads/2022/03/Strategic-Plan-2022-2026_74Y22-02-23-22.pdf.

U.S. Army Special Operations Command. *Irregular Warfare Annotated Bibliography: Assessing Revolutionary and Insurgent Strategies*. Fort Bragg, NC: U.S. Army Special Operations Command, 2011.

U.S. Department of Commerce. *The Effect of Imports of Steel on the National Security*. https://www.commerce.gov/sites/default/files/the_effect_of_imports_of_steel_on_the_national_security_-_with_redactions_-_20180111.pdf.

U.S. Department of Defense. *2022 National Defense Strategy*.

———. "Department of Defense Releases the President's Fiscal Year 2024 Defense Budget." https://www.defense.gov/News/Releases/Release/Article/3326875/department-of-defense-releases-the-presidents-fiscal-year-2024-defense-budget/.

———. *Irregular Warfare (IW) Joint Operating Concept (JOC)*. September 11, 2007. https://www.jcs.mil/Portals/36/Documents/Doctrine/concepts/joc_iw_v1.pdf?ver=2017-12-28-162020-260.

———. *Military and Security Developments Involving The People's Republic of China*. Annual Report to Congress, 2022. https://media.defense.gov/2022/Nov/29/2003122279/-1/-1/1/2022-MILITARY-AND-SECURITY-DEVELOPMENTS-INVOLVING-THE-PEOPLES-REPUBLIC-OF-CHINA.PDF.

———. *Summary of the 2018 National Defense Strategy of the United States of America*. https://dod.defense.gov/Portals/1/Documents/pubs/2018-National-Defense-Strategy-Summary.pdf.

———. *Summary of the Irregular Warfare Annex to the National Defense Strategy*. https://media.defense.gov/2020/Oct/02/2002510472/-1/-1/0/Irregular-Warfare-Annex-to-the-National-Defense-Strategy-Summary.PDF.

———. *Quadrennial Defense Review Report*. https://dod.defense.gov/Portals/1/features/defenseReviews/QDR/QDR_as_of_29JAN10_1600.pdf.

———. "United States Government Compendium of Interagency and Associated Terms." https://www.jcs.mil/Portals/36/Documents/Doctrine/dictionary/repository/usg_compendium.pdf?ver=2019-11-04-174229-423.

U.S. Department of Justice. "Deputy Assistant Attorney General Adam S. Hickey Testifies Before Senate Judiciary Committee at Hearing Titled "Election Interference: Ensuring Law Enforcement Is Equipped to Target Those Seeking to Do Harm." June 12, 2018. https://www.justice.gov/opa/speech/deputy-assistant-attorney-general-adam-s-hickey-testifies-senate-judiciary-committee.

———. "Two Iranian Nationals Charged for Cyber-Enabled Disinformation and Threat Campaign Designed to Influence the 2020 U.S. Presidential Election." Press Release, November 18, 2021. https://www.justice.gov/opa/pr/two-iranian-nationals-charged-cyber-enabled-disinformation-and-threat-campaign-designed.

U.S. Department of State. "FY 2023 International Affairs Budget." https://www.state.gov/fy-2023-international-affairs-budget/.

———. "Department Of State Learning Agenda 2022 2026 (2)." June 17, 2022. https://www.state.gov/plans-performance-budget/department-of-state-learning-agenda-2022-2026-2/.

———. "Mission & Vision." https://www.state.gov/bureaus-offices/under-secretary-for-public-diplomacy-and-public-affairs/global-engagement-center/.

U.S. Department of Treasury. "Treasury Sanctions Russian Intelligence Officers Supervising Election Influence Operations in the United States and Around the World." Press Release, June 23, 2023. https://home.treasury.gov/news/press-releases/jy1572.;

U.S. European Command Public Affairs. "USEUCOM State Partnership Program Supports Chicago COVID-19 Response with Polish Military Medical Team." April 24, 2020. https://www.eucom.mil/article/40274/useucom-state-partnership-program-supports-ch.

U.S. Joint Chiefs of Staff. *Joint Concept for Competing*. https://s3.documentcloud.org/documents/23698400/20230213-joint-concept-for-competing-signed.pdf.

———. *Information in joint operations* (JP 3-04), 2022.

U.S. Naval War College. "Center on Irregular Warfare & Armed Groups." https://usnwc.edu/Research-and-Wargaming/Research-Centers/Center-on-Irregular-Warfare-and-Armed-Groups.

Ucko, David H. "Learning Difficulties: The US Way of Irregular Warfare." *Sicherheit und Frieden (S+F)* 36, no. 1 (2018): 21–26.

———, and Thomas A. Marks. *Crafting Strategy for Irregular Warfare: A Framework for Analysis and Action*. Washington, DC: National Defense University Press, 2022.

———. "Redefining Irregular Warfare: Legitimacy, Coercion and Power." *Modern War Institute* (blog), October 18, 2022. https://mwi.westpoint.edu/redefining-irregular-warfare-legitimacy-coercion-and-power/.

———. "Systems Failure: The US Way of Irregular Warfare," *Small Wars & Insurgencies* 30, no. 1, (January 2, 2019): 223–254.

Ulrich, Philip. "Overwhelming Force—A Persistent Concept in U.S. Military Thinking." Copenhagen: Royal Danish Defense College, 2013.

UN General Assembly. "The promotion, protection, and enjoyment of human rights on the Internet." Human Rights Council, Forty-Seventh Session, 21 June-13 July 2021. A/HRC/47/L.22 https://documents-dds-ny.un.org/doc/UNDOC/LTD/G21/173/56/PDF/G2117356.pdf?OpenElement.

Unal, Beyza, Yasmin Afina, and Patricia Lewis, eds. *Perspectives on Nuclear Deterrence in the 21st Century*. London: Chatham House, 2020.

United Nations Office for Disaster Risk Reduction. " UNDRR Homepage." https://www.undrr.org/.

United Nations Security Council. "United Nations Security Council Consolidated List." https://www.un.org/securitycouncil/content/un-sc-consolidated-list.

V., Anan., "Revisiting the Discourse on Strategic Culture: An Assessment of the Conceptual Debates." *Strategic Analysis* 44, no. 3, (May 3, 2020): 193–207.

Van Atta, Richard H., Michael J. Lippitz, Jasper C. Lupo, Rob Mahoney, and Jack H. Nunn. "Transformation and Transition: DARPA's Role in Fostering an Emerging Revolution in Military Affairs." Washington, DC: Institute for Defense Analyses, 2003. https://apps.dtic.mil/sti/pdfs/ADA437248.pdf.

Van Metre, Lauren. "Fragility and Resistance." Washington, DC: United States Institute for Peace, September 20, 2016. https://www.usip.org/publications/2016/09/fragility-and-resilience.

Vahlsing, Candace. "Quantfying Risks to the Federal Budget from Climate Change." The White House Briefing Room, April 4, 2022. https://www.whitehouse.gov/omb/briefing-room/2022/04/04/quantifying-risks-to-the-federal-budget-from-climate-change/.

Voice of America. "VOA's Mission in the 1960s and 1970s." last updated February 13, 2007, https://www.insidevoa.com/a/a-13-34-20 07-mission-in-the-1960s-and-1970s-111602674/177528.html.

Wagnild, G. M., and H.M. Young. "Resilience Scale (RS). " Database record. APA PsycTests. (1988), https://doi.org/10.1037/t07521-000.

Walentek, Dawid. "Economic Peace Revisited: Coercion in a Liberal Order." Amsterdam, The Netherlands: Amsterdam Institute for Social Science Research, 2020. https://preprints.apsanet.org/engage/api-gateway/apsa/assets/orp/resource/item/5e7a5bb27fcf2a001940cdc0/original/economic-peace-revisited-coercion-in-a-liberal-order.pdf.

Walker, Vivian S., and Sonya Finley, eds. *Teaching Public Diplomacy and The Information Instruments of Power in a Complex Media Environment*. Washington, DC: Advisory Commission on Public Diplomacy, August 2020.

Walklate, Sandra, Ross McGarry, and Gabe Mythen. "Searching for Resilience: A Conceptual Excavation." *Armed Forces & Society* 40, no. 3 (2014): 408–427. https://www.jstor.org/stable/48609333.

Walmsley, Terrie, Adam Rose, Richard John, Dan Wei, Jakub P. Hlavka, Juan Machado, and Katie Byrd. "Macroeconomic Consequences of the

COVID-19 Pandemic." *Economic Modelling* 120, March 2023. https://doi.org/10.1016/j.econmod.2022.106147.

Walpole, H. Emily, Jarrod Loerzel, and Maria Dillard. "A Review of Community Resilience Frameworks and Assessment Tools: An Annotated Bibliography." Gaithersburg, MD: National Institute of Standards and Technology, August 12, 2021. https://doi.org/10.6028/NIST.TN.2172.

Walt, Stephen M. "America's Too Scared of a Multipolar World." *Foreign Policy*, March 7, 2023. https://foreignpolicy.com/2023/03/07/america-is-too-scared-of-the-multipolar-world/.

Waltz, Kenneth N. *Man, the State, and War.* New York: Columbia University Press, 1959.

———. *Theory of International Politics.* Long Grove, IL: Waveland Press, 1979.

Wang, Christoph Nedopil. "China Belt and Road Initiative (BRI) Investment Report 2022." Green Finance & Development Center, February 3, 2023. https://greenfdc.org/china-belt-and-road-initiative-bri-investment-report-2022/;

Wang, Zheng. *Never Forget National Humiliation.* New York: Columbia University Press, 2014.

Wanless, Alicia, and James Pamment. "How Do You Define a Problem Like Influence?" *Journal of Information Warfare* 18, no. 3, (2019): 1–14. https://carnegieendowment.org/files/2020-How_do_you_define_a_problem_like_influence.pdf.

Ward, Megan, Shannon Pierson and Jessica Beyer. "Formative Battles: Cold War Disinformation Campaigns and Mitigation Strategies." Washington, DC: Wilson Center, August 2019. https://www.wilsoncenter.org/sites/default/files/media/documents/publication/cold_war_disinformation_campaign.pdf.

Wardle, Claire. "Information Disorder: The Essential Glossary." Harvard Kennedy School Shorenstein Center on Media, Politics and Public Policy, 2018. https://firstdraftnews.org/wp-content/uploads/2018/07/infoDisorder_glossary.pdf.

———, and Hossein Derakhshan. "Information Disorder: Toward an Interdisciplinary Framework for Research and Policy Making." Council of Europe, 2017.

Warren, Aiden, and Philip M. Baxter. *Nuclear Modernization in the 21st Century*. London: Routledge, 2020.

Washington Post Editorial Board. "We Must Learn the Lessons of the Pandemic. A Bipartisan Commission Can Help with That." *The Washington Post*, August 18, 2020. https://www.washingtonpost.com/opinions/we-must-learn-the-lessons-of-the-pandemic-a-bipartisan-commission-can-help-with-that/2020/08/18/ff5767dc-8a34-11ea-ac8a-fe9b8088e101_story.html.

Wasser, Becca, and Stacie Pettyjohn. "Why the Pentagon Should Abandon 'Strategic Competition.'" *Foreign Policy*, October 19, 2021. https://foreignpolicy.com/2021/10/19/2022-us-nds-national-defense-strategy-strategic-competition/.

Watson, Ben, and Patrick Tucker. "These Are the US Military Bases Most Threatened by Climate Change.," *Defense One*, June 13, 2019. https://www.defenseone.com/threats/2019/06/these-are-us-military-bases-most-threatened-climate-change/157689/.

Way, Lucan. "The Real Causes of Color Revolutions." *Journal of Democracy* 19, no. 3 (2008): 55–69. https://doi.org/10.1353/jod.0.0017.

Weatherford, M. Stephen. "Measuring Political Legitimacy." *American Political Science Review* 86, no. 1 (March 1992): 149–166. https://doi.org/10.2307/1964021.

Webster, Andrew. "#Reviewing Three Dangerous Men." *The Strategy Bridge*, February 23, 2022. https://thestrategybridge.org/the-bridge/2022/2/23/reviewing-three-dangerous-men.

Weisbrot, Mark, and Jeffrey Sachs. "Economic Sanctions as Collective Punishment: The Case of Venezuela." *Center for Economic and Policy Research*, April 2019. https://cepr.net/report/economic-sanctions-as-collective-punishment-the-case-of-venezuela/.

Weiss, Linda. "Re-Emergence of Great Power Conflict and US Statecraft." *World Trade Review* 20 (2021): 152–168. https://doi.org/10.1017/S1474745620000567.

Bibliography

Weismann, Mikael, "Hybrid Warfare and Hybrid Threats Today and Tomorrow: Towards an Analytical Framework." *Journal on Baltic Security* 5, no. 1, (2019): 17–26.

Westad, Odd Arne. "Legacies of the Past." In *China and the World*, edited by David Shambaugh, 25–36. New York: Oxford University Press, 2020. https://doi.org/10.1093/oso/9780190062316.003.0002

White House. "A Declaration for the Future of the internet." April 2022. https://www.whitehouse.gov/wp-content/uploads/2022/04/Declaration-for-the-Future-for-the-internet_Launch-Event-Signing-Version_FINAL.pdf.

———. "FACT SHEET: Biden-Harris Administration Launches Year of Evidence for Action to Fortify and Expand Evidence-Based Policymaking." April 7, 2022. https://www.whitehouse.gov/ostp/news-updates/2022/04/07/fact-sheet-biden-harris-administration-launches-year-of-evidence-for-action-to-fortify-and-expand-evidence-based-policymaking/;

———. "International Strategy for Cyberspace: Prosperity, Security, and Openess in a Networked World." May 2011. https://nsarchive.gwu.edu/document/20843-04.

———. "National Security Decision Directive 32." April 1982. https://nsarchive.gwu.edu/document/20310-national-security-archive-doc-25-national.

———. "National Security Decision Directive 75." January 1983. https://www.reaganlibrary.gov/public/archives/reference/scanned-nsdds/nsdd75.pdf.

———. *2015 National Security Strategy of the United States of America*. February 2015. https://obamawhitehouse.archives.gov/sites/default/files/docs/2015_national_security_strategy_2.pdf

———. *2017 National Security Strategy of the United States of America*. December 2017. https://trumpwhitehouse.archives.gov/wp-content/uploads/2017/12/NSS-Final-12-18-2017-0905.pdf.

———. *2022 National Security Strategy*. October 2022. https://www.whitehouse.gov/wp-content/uploads/2022/10/Biden-Harris-Administrations-National-Security-Strategy-10.2022.pdf.

———. "Presidential Policy Directive -- Critical Infrastructure Security and Resilience." February 12, 2013. https://obamawhitehouse.archives.gov/the-press-office/2013/02/12/presidential-policy-directive-critical-infrastructure-security-and-resil.

White, Jeffrey. "Some Thoughts on Irregular Warfare. " *Studies in Intelligence* 39, no. 5, (1996): 51–59.

White, Steven. "Economic Performance and Communist Legitimacy." *World Politics* 38, No. 3 (April 1986): 462–482.

Wigell, Mikael. "Conceptualizing Regional Powers' Geoeconomic Strategies: Neo-Imperialism, Neo-Mercantilism, Hegemony, and Liberal Institutionalism." *Asia Europe Journal* 14, no. 2 (June 1, 2016): 135–151. https://doi.org/10.1007/s10308-015-0442-x.

———, and Vihma Antto. "Geopolitics versus Geoeconomics: The Case of Russia's Changing Geostrategy and Its Effects on the EU," *International Affairs* 92, no. 3, (May 2016): 605–627.

———, Sören Scholvin, and Mika Aaltola, eds. *Geo-Economics and Power Politics in the 21st Century.* London, Routledge: 2018 https://doi.org/10.4324/9781351172288.

Williams, Heather J., and Caitlin McCulloch. "Truth Decay and National Security." *The RAND Blog* (blog), August 1, 2023.https://www.rand.org/blog/2023/08/truth-decay-and-national-security.html.

Wilson, Clay. "Cyber Threats to Critical Information Infrastructure." In *Cyberterrorism: Understanding, Assessment and Response*, edited by Thomas M Chen, Lee Jarvis, and Stuart Macdonald, 123-140. London: Springer-Swansea University, 2014.

Wilson, Ernest J. "Hard Power, Soft Power, Smart Power." *The Annals of the American Academy of Political and Social Science* 616, (2008): 110–124.

Wishnick, Elizabeth. "Xi Jinping's Vision of a Resilient China." *CNA In Depth* (blog), January 13, 2023. https://www.cna.org/our-media/indepth/2023/01/xi-jinpings-vision-of-a-resilient-china.

Wolfley, Kyle J. *Military Statecraft and the Rise of Shaping in World Politics.* Lanham: Rowman & Littlefield Publishers, 2021.

———. "Military Statecraft and the Use of Multinational Exercises in World Politics." *Foreign Policy Analysis* 17, Issue 2 (April 2021): 1–24. https://doi.org/10.1093/fpa/oraa022.

Wong, Audrye. "The Age of Informational Statecraft." *Project Syndicate*, June 22, 2022.

Woo, Stu, and Daniel Michaels. "China Buys Friends With Ports and Roads. Now the U.S. Is Trying to Compete." *Wall Street Journal*, July 15, 2021. https://www.wsj.com/articles/china-buys-friends-with-ports-and-roads-now-the-u-s-is-trying-to-compete-11626363239.

Wood, Robert J. *Information Engineering: The Foundation of Information Warfare*. Maxwell Air Force Base, AL: Air War College, April 1995. https://apps.dtic.mil/sti/pdfs/ADA329024.pdf.

Woodward, Susan L. "Can We Measure Legitimacy?" *Südosteuropa* 60, H. 4, S. (2012): 470–482.

World Bank. "Climate Explainer: Food Security and Climate Change." October 17, 2022. https://www.worldbank.org/en/news/feature/2022/10/17/what-you-need-to-know-about-food-security-and-climate-change.

———. "Worldwide Governance Indicators." https://www.govindicators.org/.

World Health Organization. "Fighting COVID-19 could cost 500 times as much as pandemic prevention measures." August 28, 2020. https://extranet.who.int/sph/fighting-covid-19-could-cost-500-times-much-pandemic-prevention-measures.

———. The General Agreement on Tariffs and Trade, 1947, Article XXI. https://www.wto.org/english/res_e/booksp_e/gatt_ai_e/art21_e.pdf.

———. "Infodemic." https://www.who.int/health-topics/infodemic#tab=tab_1.

Worley, D. Robert. *Orchestrating the Instruments of Power: A Critical Examination of the U.S. National Security System*. Lincoln, NE: Potomac Books, 2015.

Wrong, Dennis H. *Power: Its Forms, Bases and Uses*. New York, NY: Routledge, 2017.

Wuthnow, Joel. "Just Another Paper Tiger? Chinese Perspectives on the U.S. Indo-Pacific Strategy." *Strategic Forum* 305, National Defense University, June 2020.

Xinmin, Sui. "India's Strategic Culture and Model of International Behavior." *China International Studies* 45, (2014): 139–162.

Yablokov, Ilya. "Russian Disinformation Finds Fertile Ground in the West." *Nature Human Behavior* 6, No. 6, (June 2022): 766–767. https://www.nature.com/articles/s41562-022-01399-3.

Yang, Mary. "Can the CHIPS Act Put the U.S. Back in the Game?" *Foreign Policy,* August 26, 2022. https://foreignpolicy.com/2022/08/26/chips-act-us-china-semiconductor-tech-taiwan/.

Yoshihara, Toshi, and James Holmes. *Red Star over the Pacific.* 2nd Edition. Annapolis, MD: Naval Institute Press, 2018.

Yun, Seong-Hun. "An Overdue Critical Look at Soft Power Measurement: The Construct Validity of the Soft Power 30 in Focus." *Journal of International and Area Studies* 25, no. 2 (2018): 1–20.

Zakaria, Fareed. "The dollar is our superpower, and Russia and China are threatening it." *Washington Post Opinion,* March 24, 2023. https://www.washingtonpost.com/opinions/2023/03/24/us-dollar-strength-russia-china/.

———. "The Reagan Strategy of Containment." *Political Science Quarterly* 105, no. 3 (Autumn 1990): 373–395.

Zakem, Vera, Paul Saunders, and Daniel Antoun. "Mobilizing Compatriots: Russia's Strategy, Tactics, and Influence in the Former Soviet Union." *Center for Naval Analysis,* November 2015.

Zaman, Rashed Uz. "Kautilya: The Indian Strategic Thinker and Indian Strategic Culture." *Comparative Strategy* 25, no. 3, (June 1, 2006): 231–247.

Zarate, Juan Carlos. *Treasury's War: The Unleashing of a New Era of Financial Warfare.* PublicAffairs, 2013.

Zebrowski, Chris. "The nature of resilience." *Resilience* 1, no. 3, (2013): 159–173. DOI: 10.1080/21693293.2013.804672.

Zegart, Amy, and Michael Morell. "Spies, Lies, and Algorithms, Why U.S. Intelligence Agencies Must Adapt or Fail." *Foreign Affairs*, May/June 2019. https://www.foreignaffairs.com/united-states/spies-lies-and-algorithms

Zekulić, Vlasta, Christopher Godwin, and Jennifer Cole. "Reinvigorating Civil–Military Relationships in Building National Resilience." *The RUSI Journal* 162, no. 4, (2017): 30–38, https://pure.royalholloway.ac.uk/ws/portalfiles/portal/30776755/CIV_MIL_Resiliene_JC.pdf.

Zhang, Liang, Andrew J. Nathan, Perry Link, and Orville Schell. *The Tiananmen Papers*. New York, NY: PublicAffairs, 2002.

Zhang, Wanfa. "Has Beijing Started to Bare Its Teeth? China's Tapping of Soft Power Revisited." *Asian Perspective* 36, no. 4, (2012): 615–639.

Zimmerling, Ruth. *Influence and Power: Variations on a Messy Theme*. Dordrecht: Springer, 2005.

Zimmerman, S. Rebecca, Kimberly Jackson, Natasha Lander, Colin Roberts, Dan Madden, and Rebeca Orrie. *Movement and Maneuver: Culture and the Competition for Influence Among the U.S. Military Services*. Santa Monica, CA: RAND, 2019.

Zolli, Andrew, and Ann Marie Healy. *Resilience: Why Things Bounce Back*. New York, NY: Simon & Schuster, 2013.

Index

active measures, xi, 48, 63–64, 154
Active Measures Working Group, 154
"advise and assist" missions. *See* security cooperation
Afghanistan, 1, 78, 80, 108, 156, 195
Africa, 51, 53–57, 64, 105, 197
age of crises, 2, 17, 188, 219–220
Alexis de Tocqueville, 33
Almond, Gabriel A., 25, 39
al-Qaeda, 121
American exceptionalism, 30, 33–34, 77, 82
American Revolution, 147
American strategic culture, 5, 15, 21–22, 26–29, 32, 35, 40, 82, 90, 93, 227
   action bias, 31
   disadvantage, 37
   and irregular warfare, 30, 43–44, 219, 222–223
   negative aspects, 224
   positive aspects, 224
   versus China and Russia, 61
anti-access/area denial (A2/AD), xi, 91
Arab Spring, 171
Aristide, Jean-Bertrand, 154
arms control, xi, 76, 97, 104, 107
arms control, xi, 76, 97, 104, 107
Art of War, The. See Sun Tzu
Arthashastra, The. See Kautilya
artificial intelligence (AI), 3, 59, 121, 126, 131, 135, 158, 228
Ashford, Emma, 17, 85, 215,

Ashford, Emma (*continued*), 230–231
Asia Power Index. *See* indexes
Asian Barometer Survey, 202, 212
Asian Infrastructure Investment Bank, 54, 65
assessment, 16, 155, 178–180, 208, 210, 224
   culture of, 209
asset freezes, 124–125
Austin, Lloyd, 216–217
Australia, 102, 118, 136, 173, 188–189
Australia-UK-US (AUKUS), 102
Austria, 58
Axis of Evil, 35
Axis of Resistance, xx

Baldwin, David A., 17–19, 113, 136
Bargués, Pol, 173, 189
Beckley, Michael, 201, 211
Belt and Road Initiative (BRI) . *See* China, 51, 54, 65, 119, 126, 217, 225, 230
Biden administration, 8, 68, 112, 133, 157, 209, 217
bilateral agreements, 222, 229
Bismarck, Otto von, 30
blacklist, 124–125
Blechman, Barry M., 94, 110–112
blind spots, 25, 31, 33
bloodletting. *See* military tools
Bonnano, George A., 176, 189
Boot, Max, 73, 86–87
Borghard, Erica D., 167, 185, 188,

Borghard, Erica D. (*continued*), 190–191
Boyle, Michael J., 28–29, 33, 39–40
Brands, Hal, 70, 76, 78, 83–84, 86–88
Bretton Woods, 54, 113, 117, 134, 136, 215
Brice, Benjamin, 33, 40
BRICS, 54, 56
Broad Agency Announcements (BAAs), 208
Broadcom, 126, 138
Brooks, Rosa, 93, 110, 230
Bureau of Conflict and Stabilization Operations, xix
bureaucracy, 17, 26, 45, 71, 73, 78, 81, 90, 92, 108, 134, 225
Bush administration, 28
Bush, George W., 28, 35, 40

capitalism, 68, 114, 164
 data capitalism, 151
Carr, Edward Hallet, 150, 163
Carter administration, 77
 sanctions, 78
Carter, Jimmy, 77–78
Central Asia, 51, 64
Central Intelligence Agency (CIA), 73
China, xx, 2–5, 8, 17, 64, 66, 80, 82, 85, 87, 100–102, 106, 111, 118, 120–122, 130–131, 135–138, 146, 151, 155, 157, 160, 164, 188, 191, 198, 206, 211, 215, 218, 220, 229, 231
 advantage over the US, 60
 Belt and Road Initiative (BRI), 51, 54, 65, 119, 126, 217, 225, 230
 bureaucracy, 45
 Century of Humiliation, 45–47

China (*continued*)
 China Dream, 45, 63
 China Global Television Network, 57
 Chinese Communist Party (CCP), 45–47, 50–51, 56–57, 59, 68
 Chinese philosophy, 45
 civil society, 45
 Confucianism, 45
 Confucius Institutes, 57, 65
 control of information, 56, 59, 149–150
 co-opting information, 57
 Cross-Border Interbank Payment System, 55
 Cultural Revolution, 45
 demands positive portrayal, 81
 disadvantage, 61
 economic influence, 43–44, 46, 53–56, 60–61, 114, 119, 126, 129, 134, 170, 201–202, 216–217
 economic resilience, 59, 61
 economic statecraft, 46, 53–55, 114, 126, 134, 225
 economy, 46
 history and geography, 44–45
 investment in infrastructure worldwide, 54
 irregular warfare, 9, 11, 43–47, 55–56, 60–62, 91, 141, 158, 170, 183, 202, 214, 217, 219, 224
 Legalism, 45
 media and non-governmental organizations, 149
 military statecraft, 50–52, 104, 108
 National Rejuvenation, 45
 national resilience, 44, 58–59,

China
  national resilience (*continued*), 170, 216
  People's Armed Forces Maritime Militia (PAFMM), 50, 105
  People's Liberation Army (PLA), 50–51, 57
  political structure, 45
  Renminbi (RMB), 55, 129
  revisionist territorial claims, 50
  sovereignty, 46
  strategic culture, 44–47, 61
  surveillance state, 59
  Three Warfares, xv, 56–57, 65
  Tiananmen Square protests and massacre, 45
  vulnerabilities, 45–46
  vulnerabilities, 61, 134, 217
  weaknesses, 61
  See also Xi Jinping
CHIPS and Science Act, 135
Clark, William, 78, 230
Clausewitz, Carl von, 4, 17, 25, 36, 64, 69, 89, 110
Cleveland, Charles T., 80, 87, 230
climate change, 1–4, 6, 11, 13–14, 168–169, 181, 184, 186, 207, 219–220
coercion, xvii, 110–112, 191, 230
  economic coercion, 5, 10–11, 61, 134, 170, 183
cognition, 28, 144
Cold War, xii, xvii, 2, 6, 15, 46, 48, 68, 78, 83, 90–92, 106, 108–109, 121, 123, 127–128, 135, 141, 154, 156–157, 182, 215, 218, 222, 224, 227
  Cold War 2.0, 217
  political warfare, 11, 67, 69–73, 76–77, 79–82, 84, 187, 195
  post-Cold War, 1, 105, 213

collective defense, 186, 222
collective resilience. *See* resilience
Committee on Foreign Investment in the United States (CFIUS), 126, 138
communism, 68, 73–74, 77
Communities Advancing Resilience Toolkit (CART). *See* resilience
community resilience. *See* resilience
compellence, xi, 93
competitive statecraft, 11–12, 224
Composite Index of National Capabilities (CINC). *See* indexes
conditionalism, xi
Confucianism. *See* China
Congress, xviii, 59, 71, 74–75, 86, 126, 133, 136, 155, 188, 209, 216, 230
Congressional Research Service, 209, 230
containment, xii, 11, 69–70, 72–74, 76, 85–87, 195
conventional forces, xii, 104, 109
conventional warfare, xviii, 10
Cooper, Evan, 17, 85, 215, 230–231
Cordesman, Anthony H., 80, 82, 87
Correlates of War (COW). *See* indexes
Countering Foreign Influence Task Force (CFITF), 159
Countering Foreign Propaganda and Disinformation Act, 159
counterterrorism. *See* terrorism
COVID-19, 2, 103, 118, 134, 136, 143, 165, 170–171, 173, 183, 185, 189–191, 219, 221
Crimea, 2, 102

Cross-Border Interbank Payment
  System. See China
culture
  American cross-cultural fluency,
    32
  differences, 22, 32
  hierarchy, 22, 25, 27, 45
  subculture, xii–xiii, xv, 22,
    25–27, 38
currency, 54–55, 124, 129–130, 138
cyberattacks, 11, 68, 81, 83, 97, 183,
  186, 220, 223, 229
CYBERCOM, 157
cybersecurity, 159, 167–168
Cybersecurity and Infrastructure
  Security Agency (CISA), 159

data capitalism. See capitalism
Declaration for the Future of the
  Internet, 157, 165
Declaration of Independence, 32, 40
Defense Advanced Research
  Projects Agency (DARPA), 135
defense innovation, xii, 109
Defense Innovation Unit, 109
defense institution building (DIB).
  See security cooperation
defense sector, 51, 53, 90, 131
defense spending, 76, 109, 118, 219
democracy, xiii, xv, 2, 16, 26, 28, 30,
  33–34, 40, 58, 61, 63, 73–74, 77,
  79, 119, 123, 142–143, 145, 148,
  152, 154, 160, 174, 180, 187, 189,
  199, 202, 212, 224
  global legitimacy of, 155
Department of Commerce, 121, 137
Department of Defense (DOD),
  xvii–xx, 4, 6, 9–10, 17–18, 27, 35,
  63–64, 66, 72, 91–92, 104, 106,
  108, 110–111, 131, 135, 148, 211,
  216–217, 220, 226

Department of Defense (DOD)
  (continued)
  DOD's innovation ecosystem,
    109
Department of Homeland Security
  (DHS), 159, 165, 178
Department of Justice, 155, 164–165
Department of State (DOS), 35, 69,
  75–76, 108, 132, 155, 164, 212,
  217, 227, 230
deterrence, xii, xx, 93, 97, 107, 182,
  191, 223, 230
  to allies, 98
  integrated deterrence, xviii, 106,
    217, 220
  nuclear deterrence, 39, 90, 104
diplomacy, 12, 17–18, 27, 31, 36, 51,
  64–65, 89, 93, 100, 136, 144, 170
disasters, 99, 168, 178
  natural, 2–3, 91, 169, 183–185,
    219, 228–229
disinformation, xi–xiii, 2–4, 15,
  36, 48, 57–58, 61, 65–66, 83,
  97, 142–143, 147, 151, 154–157,
  159–162, 164–165, 170, 177, 186,
  188, 224
  best defense against, 158
  fake news, 144, 189
Disinformation Governance Board,
  159
domestic policies, 12, 115, 118, 120,
  127–132, 203
Doran, Michael, 73, 86–87
drone, 131
Drucker, Peter F., 193, 211

economic assistance, 4, 119, 124,
  194, 206, 225
  US, 126–127
Economic Competition Strategy,
  132

economic influence
  China, 43–44, 46, 53–56, 60–61, 114, 119, 126, 129, 134, 170, 201–202, 216–217
economic interdependence, 68, 114–115
economic legitimacy, 120
economic liberalism, xii, 117
economic policies, 59, 74, 115, 120, 144
economic power, 27, 118–119, 124, 127–128, 132–133, 203, 206, 215, 217
economic statecraft, 12, 15, 17–19, 50, 113–116, 126, 132–138, 225
  Chinese, 43, 46, 53–55
  conditional approach, 117
  Russian, 43, 55–56
economic tools, 203
  capital-based, 115, 118, 120, 123, 127, 132
  embargoes, 78, 120–122, 137
  financial tools, 123, 127
  inducements, 119
  sanctions, 3, 119–121, 124, 130, 134, 220, 225
  trade-based, 115, 118, 120–121, 123–124
  US, 5, 11, 60–61, 114, 117, 120, 124, 127, 130, 132–135, 170, 220, 222, 224
Egel, Daniel, 132, 139
Egypt, 107
energy, 56, 65, 77–78, 87, 105, 128, 130, 138, 172, 180, 197, 219, 227, 229
Europe, 58, 76, 105, 107, 110, 127, 130, 136, 164, 191, 216, 220, 230–231
  Eastern Europe, 47, 52, 143
  Western Europe, 73–74

expenditure, 2, 6, 40, 59, 65, 71, 76, 94, 105, 109, 136, 171, 197, 202, 211, 217
  future US, 107
  imbalance in American security spending, 219
  on conventional military power, 216
  on national resilience efforts, 220
  on non-traditional threats, 184
  projected US, 168
  spending priorities, 107
  worldwide military expenditures, 118
Export Control Act, 121
export controls, 121, 137
Export Reform Act, 121

fake news. *See* disinformation
Federal Advisory Committee Act (FACA), 156
Federal Bureau of Investigation (FBI), 154–155, 159, 164
Filloux, Frederic, 142, 162
Financial Action Task Force (FATF), 113–114
Five 5G technology, 126, 131, 135
force posture, xii, 50, 105, 217
Foreign Bilateral Influence Capacity (FBIC). *See* indexes
foreign direct investment (FDI), 115
foreign exchange, 60, 129
Foreign Influence and Interference Branch (FIIB), 159
Foreign Influence Task Force (FITF), 155, 159
Foreign Influence Task Force (FITF), 155, 159
Foreign Intelligence Surveillance Act (FISA), 156, 165

foreign internal defense (FID), 10, 168, 188
Foreign Investment Risk Review Modernization Act, 126
foreign malign influence, 144, 151, 155, 157, 159, 162
Foreign Malign Influence Center (FMIC), 159, 162
Foreign Military Financing (FMF), 101, 106–107
Foreign Military Sales (FMS), 12, 101, 106–108
foreign policy, xii, xiv, 1, 4, 8, 26, 28, 31, 39–40, 44, 50, 63–64, 92–93, 136, 150, 194, 208–209, 231
  top five American concerns, 220
Fragile States Index (FSI). *See* indexes
frameworks, 28, 174, 179, 181, 184, 194, 196, 200–202, 204, 207–208, 228
Freedom House, 199, 211
freedom of navigation operation (FONOPS), xii
freedom of speech, 161
freedom of the press. *See* journalism
Friedman, Uri, 171, 188, 230
Fund for Peace, 58, 65, 176, 189, 199, 211

Gaddis, John Lewis, 73–74, 86–87
Galeotti, Mark, 37, 41, 63–64
General Agreement on Tariffs and Trade (GATT), 113, 122, 137
Georgia, 52
Germany, 1, 64, 98
Gershaneck, Kerry K., 81, 87
Giske, Mathilde Tomine Eriksdatter, 2, 17, 171, 188

Global Engagement Center (GEC), 155, 159
Global Magnitsky Human Rights Accountability Act of 2016, 125
Global War on Terror, 1
Goodhart's law, xii, 207, 212
Google, 131
Gorbachev, Mikhail, 149
Government Accountability Office, 209
Government in the Sunshine Act, 156
grand strategy, xx, 2, 4, 15–16, 65–67, 69, 71–72, 79, 81, 84–85, 93, 168, 184, 188, 190–191, 193, 207, 213–215, 217–218, 220, 223, 225
  framework for, 6, 221, 227
  purview of the DOD, 6, 92
  whole-of-society, 5, 135, 145, 147–148, 159, 222, 226, 228–229
gray zone, xii, xvii–xviii, 5, 9, 18, 35, 37, 80, 82, 87
  integrated approach, 227
  and irregular warfare, 6, 15, 69, 193, 214–215, 221, 223, 226–227
  and resilience, 6, 16, 85, 168, 184, 190–191, 207, 220–227, 229
Gray, Colin S., 17–18, 23–24, 28–29, 31–32, 36, 39–41, 87
Great Power Competition (GPC), 7–8, 17–18
Gross Domestic Product (GDP), 53–54, 133, 136, 171, 185, 197, 201, 203
Gui, Yongtao, 183, 191
Gulf War, 104

Haiti, 154

Index

Hamas, xx, 216
Hamiltonianism, xii, 26, 28, 221
hard power, 6, 186, 197, 218, 226
Healy, Ann Marie, 169, 188–189
hegemony, 33, 213, 218
Hoffman, Frank, 39, 70, 82, 86–87
holistic approach, 4, 6, 8, 10, 17, 50, 59, 92, 106, 132, 167, 208, 222, 225, 227, 229
Hollywood, 81
Huntington, Samuel P., 34, 40
Hussein, Saddam, 34
hybrid warfare, xii, 5

indexes, 194, 196, 202, 204, 207–208, 228
    Asia Power Index, 198, 203, 211
    Composite Index of National Capabilities (CINC), 197, 201
    Correlates of War (COW), 197, 201, 211
    Foreign Bilateral Influence Capacity (FBIC), 197–198
    Fragile States Index (FSI), 199–200, 211, 223, 231
    Resilience Index (SRI), 58–59, 65, 176, 189, 199, 211
    Worldwide Governance Indicators, 198, 211
India, 36, 41, 54, 129
individual resilience. *See* resilience
Indo-Pacific, 51, 107, 198, 216
    Indo-Pacific Economic Framework, 183
industrial policy, 60, 128, 131–132, 138
influence, xi, xv, xvii, 4–6, 8, 10–11, 13, 15–16, 18–19, 23, 34, 43, 45–48, 50–58, 60–62, 64, 70, 77, 81, 91, 94, 96, 98–100, 105, 108–110, 113–114, 117, 119–121,

influence (*continued*), 124, 126–133, 135, 142, 144, 148, 150–152, 155–159, 161, 169–170, 172–173, 176, 186, 189, 193, 195–196, 198–200, 202–204, 206–208, 211, 215–216, 218, 220, 223, 226, 229
    definitions, xiii, 14
    influence operations, xiii, 145–147, 162–164
    US influence, 44, 101, 103, 134, 147, 183, 197, 201, 217, 221
Influence and Perception Management Office (IPMO), 159
infodemic, 143, 162
information
    false, xii, xiv, 57, 60, 144, 151, 158, 161
    information authentication, 159
    information pollution, 142
    key tools, 12
    trustworthiness of, 143, 148, 151–152
    as a tool of statecraft, 145, 152, 154, 157
    freedom of information, 149, 161
    weaponized, 161
    inoculation theory, 177
    See also disinformation *and* misinformation
information and media literacy, 158
information cooperation, 161
information environment, xviii, 145, 147, 149–150, 154, 158, 160, 167
    three components, 144
information influence, 151, 162, 164
information infrastructure, 147, 156–157, 165
information legitimacy, 152
information manipulation, 12,

information manipulation
(*continued*), 151–152, 155–157
information operations, xiii, 5, 10,
58, 81, 83, 145–147, 158, 163
definition, 148
information power, 150–151
information sharing, 98, 149, 152,
154–155, 160
information statecraft, xiii, 12, 50,
145, 150, 152–153, 158–159, 161
Chinese, 43, 56–58, 141–142, 146
definition, 146
Russian, 141–142
theories of, 148
information strategy, 160, 162,
164–165
guarded openness, 149
whole-of-society, 159
information tools, 146, 151–152,
170, 203
information warfare, xiii, xx,
145–148, 162–164
campaigns, 160
infrastructure, 11, 54–55, 65, 115,
119–120, 126, 128, 130, 158–159,
178, 182, 184–185
critical, 3, 5, 72, 171–172, 179,
181, 219, 221, 225
destruction of, 12, 152
information, 147, 156–157, 165
innovation, xii, 27, 78, 104, 106, 109,
115, 128–129, 131, 135, 190, 197,
225
intelligence, 3, 38, 48, 50, 58, 66, 73,
91, 98, 102–103, 106, 147–148,
150, 155, 162, 164, 171, 188, 190
foreign intelligence, 12, 156, 165
intermediate-Range Nuclear Forces
Treaty. *See* nuclear weapons
international agreements, 91, 152,
157

International Covenant on Civil
and Political Rights (ICCPR). *See*
United Nations
International Finance Corporation
(IFC), 113, 127
International Military Education
and Training (IMET), 100, 111
International Monetary Fund (IMF),
113, 127, 138
International Professional Military
Education (IPME), 51, 100
international resilience. *See*
resilience
*International Strategy for
Cyberspace*, 149
international system, 9, 13, 27, 68,
90, 132, 161, 170, 172, 183, 194,
223, 228
bipolar, 215
multipolar, 215, 222
Internet Research Agency (IRA).
See Russia
Iran, xx, 44, 53, 105, 157
Iraq, 1, 33–34, 80, 107
irregular warfare (IW), xv,
xvii–xviii, xx, 7, 12, 15, 17–18,
30, 36, 44–45, 47, 55, 62–63, 69,
82, 93–94, 96–97, 105, 109, 117,
121, 123, 127, 132, 135, 141–142,
150, 158, 168, 172–174, 182, 184,
186–187, 189, 193, 195–197,
200–202, 204–209, 211, 214–216,
219, 222–224, 226–229
competitor approaches to, 43,
50, 199
definitions, xiii, 6, 9–11, 13, 61,
70, 167, 169, 221
binary categorization of threats
and tools, 9
and DOD, xix, 6, 9–10, 91–92,

irregular warfare (IW)
and DOD (*continued*), 217
tools, 46, 48–50, 53, 56, 60, 170, 183
violence as a necessary condition, 10
Irregular Warfare Dashboard, 195–197, 204–205, 207–209
two key principles, 206
Irregular Warfare Functional Center (IWFC), xviii
isolation, 27, 195, 197, 206
geographic, 30, 32–33
isolationism, 28, 31, 40, 220–221, 231
Israel, xx, 107, 121, 216
Italy, 58

Jacksonianism, xiii, 26, 221
Japan, 1, 98, 202
Jeffersonianism, xiii, 26, 28, 31
John Paul II (pope), 77
Johnson, Jeannie L., 24, 29, 31, 40, 63, 66, 138, 189, 212
Johnston, Alistair Iain, 24–25, 39
Joint Concept for Competing, xviii, 8, 17–18
Joint Staff, xviii, 188
Joint Warfighting Concept, xviii
Jones, Seth G., 11, 17–18, 63–65, 70, 80, 82, 84–85, 87–88, 164–165
Jordan, xx, 107, 110–111
journalism, 60–61, 151–152, 165
freedom of the press, 66, 77, 149, 160
Society for Professional Journalists, 160
junior leaders, 100–101

Kautilya, 35, 41, 89, 92, 110, 224

Kautilya (*continued*)
Arthashastra, The, 36
silent war, 36
Kennan, George F., 11, 18, 35, 41, 69–72, 75, 85–86
Long Telegram, 133
Key Leader Engagements (KLEs). *See* National Defense Strategy (NDS)
KGB. *See* Soviet Union
Klare, Michael T., 171–172, 188–189, 230
Korean soldiers in US formations (KATUSAs). *See* South Korea
Korean War, 72, 74
Kremlin Playbook. See Russia
Kremlin. *See* Russia
Krishnan, Armin, 9, 18
Kuhn, Thomas S., 28, 39

Lang, Anthony F., 28–29, 33, 39–40
language, 32, 57–58, 77, 158, 224
League of Nations, 34
legal warfare, xv, 57
Legalism. *See* China
legitimacy, xi–xiii, xv, 4–6, 8, 10–11, 13, 15–16, 19, 34, 36, 44, 46, 48, 50, 52–57, 60–62, 67, 70, 77, 91, 94, 96, 99–100, 103, 105, 109, 114, 117, 119, 121, 125, 127–128, 132–136, 142–143, 145–146, 150–152, 155–156, 160–161, 169–170, 173–174, 176, 178, 183, 186, 189, 193, 195–204, 206–208, 211–212, 216–218, 221, 223, 226, 229
definition, 14
economic legitimacy, 120
Legro, Jeffrey, 26, 39
Lend-Lease Act, 90

Liebig, Michael, 36, 41
Likert scale, 31, 207
liquified natural gas (LNG), 130
Lithuania, 56
Lowy Institute, 198, 211
Lucas, Scott, 71, 85–86, 211

Mac Thornberry National Defense Authorization Act, xviii
Mahnken, Thomas G., 26–27, 39, 70, 85–86
Maine, Henry James Sumner, 213, 230
Mansted, Katherine, 141, 162, 164–165
Mao Zedong, 45
Marks, Thomas A., 17–18, 172, 189
Marshall Plan, 73–75, 85, 127, 186, 222
Marshall, George, 75
Mazarr, Michael J., 18, 80–81, 87, 211
McCullough, Caitlin, 173
McFate, Sean, 43, 63
McMaster, H. R., 34, 40, 226, 231
Mead, Walter Russell, 26–28, 33, 39–40, 231
Mearsheimer, John J., 13, 97, 110, 211
measure of effectiveness (MOEs), xiv, 195–196, 207–208
measure of performance (MOPs), xiv, 195–196
measurement, 31, 188, 200, 207, 209–210, 227–228
  approaches, 194
  measures of performance (MOPs), xiv, 195–196, 204
  measures of effectiveness (MOEs), xiv, 195–196, 203, 206, 208

measurement (*continued*)
  of power, influence, and legitimacy, 195–196, 199
  See also frameworks, indexes, and Irregular Warfare Dashboard
mercantilism, xiv, 117
Microsoft, 131
Middle East, xx, 2, 51–53, 57, 105, 164
military culture, 5, 21–22, 27–28, 30–32, 36–37, 39–40, 43, 45, 47–48, 90, 93, 115, 223–224
  subculture, 26
military exercises, xiv, 50–51, 53, 102–103
military influence, 98, 100, 108
military information support operations (MISO), 99, 103
military legitimacy, 99, 103
military means, xiv, 3, 12, 50, 91–92, 96, 217
  non-kinetic, 94
military power, 4, 50, 52, 98, 101, 103–104, 108–109, 115, 118, 171, 201, 216, 218
  active sources, 72, 96
  latent sources, 96
  multilateral and bilateral alliances, 97
military statecraft, xiv, 12, 15, 64, 89, 91, 94–96, 99, 103, 105, 109–111, 115, 156, 225
  Chinese, 43, 50–52, 104, 108
  definition, 92
  holistic approach, 50, 106
  peacetime, 90, 93
  Russian, 43, 52–53, 108
military tools, 53, 89, 91–92, 106, 108
  bloodletting, 97

military tools (*continued*)
  categorization of, 94
  interpersonal tools, 99–101
  organizational tools, 101–103
  systemic tools, 103–105, 109
mirror imaging, 32, 34, 224
misinformation, xiv, 3, 142–144, 154, 161, 177, 223, 225–226, 228
  best defense against, 158
Mistry, Kaeten, 71, 85–86, 211
Moldova, 52
monetary and fiscal policy, 128–129
Moore's Law, 142
Morillas, Pol, 174, 189
Multilateralism, 134, 214

National Aeronautics and Space Administration (NASA), 135
National Defense Authorization Act (NDAA), xviii, 124, 159
National Defense Strategy, xviii, 7, 17, 110–111
National Defense Strategy (NDS), xviii, 7, 17, 98, 106, 110–111, 182
  Key Leader Engagements (KLEs), 92, 99–100, 107
National Endowment for Democracy, 73
National Institute of Standards and Technology (NIST), 179
National Institutes of Health (NIH), 135
National Intelligence Council, 171, 188, 190
national power, 6, 9, 11, 46, 72, 81, 136, 150, 172, 196, 201, 211, 217–218, 222, 226
national resilience. *See* resilience
National Security Agency, 157

National Security Act of 1947, 70–71, 85
National Security Decision Directive 75, 77, 86
National Security Strategy (NSS), 1, 7, 13, 17–18, 41, 77, 79, 85, 87, 132, 146, 163, 168, 185, 188, 216, 219, 230
National Technology Industrial Base, 102, 111
NATO, 47, 52–53, 61, 82, 107, 216, 220
  Article 2, 74–75
  Article 5, 74
  NATO Military Personnel Exchange Program, 101
  on resilience, 180, 182–183, 185, 188, 190–191
natural disasters. *See* disasters
natural gas, 130
Nazi, 34
Ngoun, Kimly, 174, 189
9/11 attacks, 28, 80, 85, 91, 123–124, 194
noncommissioned officers (NCOs), 100, 111
non-kinetic tools, 5, 10, 94, 106
North Korea, 44, 53, 81
  North Korean Army, 72
nuclear weapons, 22–23
  China, xx, 11, 50, 55, 91, 104
  Intermediate-Range Nuclear Forces Treaty, 104
  Iran, 105
  nuclear advantage, 104
  nuclear arms race, 104
  nuclear deterrence, 39, 90, 104
  Russia, 11, 52, 55, 64, 81, 91, 97, 104, 112, 123
Nye, Joseph S., Jr., 18, 149, 163, 211, 231

Office of Foreign Asset Management (OFAC), 124–125
Office of the Director of National Intelligence (ODNI), 155, 162, 164
　Election Threats Executive, 159
oil, 78, 121, 125, 129–130, 138, 179
Organization for Economic Co-operation and Development (OECD), 113, 189
Overseas Humanitarian, Disaster Assistance, and Civic Aid (OHDACA), 103
Owens, William A., 40, 149, 163

Palestine, xx
Panama, 119, 136
Paris Call for Trust and Security in Cyberspace, 157, 165
Pavel, Barry, 64, 112, 132, 139
Pearl Harbor attack, 28
People's Armed Forces Maritime Militia (PAFMM). *See* China
People's Liberation Army (PLA). *See* China
Poland, 56, 103
　Solidarity movement, 73, 77
Poland-Illinois SPP. *See* State Partnership Program (SPP)
policymakers, 4–5, 7, 9–10, 32, 34, 37, 43, 80, 90–91, 93, 96, 101, 106, 108–109, 117, 121, 125, 149, 194, 197–202, 204, 209, 218, 225
　understanding of geopolitical threats, 3
political culture, xiv, 24–28, 39, 45
political culture, xiv, 24–25, 39
　American, 21, 26–27, 44, 61–62, 82
　Chinese, 45

political culture (*continued*)
　Hamiltonian, 26, 28
　Jacksonian, 26
　Jeffersonian, 26, 28
　Wilsonian, 26, 28–29, 34
political power, 10, 14, 44
　power of opinion, 150
political warfare, xi, xiv, 5, 62, 65, 74, 85–88, 165, 225, 230
　authoritarian, 82
　definition, 69
　during Cold War, 11, 67, 69–73, 76–77, 79–82, 84, 187, 195
　during Reagan years, 69, 76–78
　essential tools, 83
　integrated, 75, 81–82
　terminological quagmire, 82
Posen, Barry R., 96, 110
power, xi, xiii–xiv, xvii–xviii, 3, 5–7, 15–19, 27–28, 31, 34, 40, 43–46, 48, 50–55, 58–62, 65, 68, 72, 77, 81, 83, 86–87, 91, 94, 97–99, 101, 103–105, 108–109, 114–115, 117–121, 123–124, 127–133, 135–137, 139, 142, 150–152, 155–156, 160–161, 164–165, 169–170, 172, 176, 184, 186, 193, 195, 197–200, 203–204, 206–208, 216–218, 220–223, 226, 228–229
　calculations of power, 171
　definitions, 4, 8–11, 13–14, 96, 146, 196, 201–202, 211, 215
　four primary instruments, 12
Presidential Policy Directives, 168
Prier, Jarred, 147, 162–163
Prigozhin, Yevgeny. *See* Wagner Group
private military and security companies (PMSCs), 51
private sector, xiii, 124, 129, 131,

private sector (*continued*), 135, 148, 157, 159–160, 179, 181, 208, 222, 227
Project Maven, 131
Pronk, Danny, 72, 86
propaganda, xiv, 10, 12, 36, 66, 69, 77, 79, 143–144, 147, 157, 159, 162, 164–165
property rights, xii, 115, 127–128
prosecution of nefarious actors, 158
protectionism, xiv, 114, 133, 222
psychological warfare, xv, 56–57, 70
public diplomacy, 12, 18, 144
public goods, 127–128, 133, 174, 181
public opinion, xiii, xv, 56–57, 69, 75, 143, 145, 147, 152, 172, 202
public opinion warfare, xv, 56–57
public-private partnerships, 131, 179, 227
Putin, Vladimir, 44

Qualcomm, 125–126, 138

Radio Free Europe, 73, 164
Radio Liberty, 65, 73, 164
RAND, 149
 Understanding a New Era of Strategic Competition, 201, 203
 "Understanding A New Framework for Competition," 196
Reagan administration
 political warfare campaign, 69, 76–78
 sanctions, 78
Reagan, Ronald, 76–77
realism, 36, 117
recommendations, 16, 91, 105, 114, 132, 142, 158, 184, 193–194, 208,

recommendations (*continued*), 214, 225
 engaging partners and allies, 228
 integrated approach to grand strategy, 227
 measurement framework, 227
 offensively focused measures, 229
 reframing national security, 226
 shift in mindset, 226
 whole-of-society, 5, 226, 228–229
Renminbi (RMB). *See* China
reorientation of priorities, 6
Requests for Proposals (RFPs), 208
research and development (R&D), 128, 133, 138
resilience, 3, 17, 85, 197, 227, 229
 anticipatory and reactive, 169
 authoritarian, 43–44, 174, 182, 187, 189
 Brief Resilience Scale, 176, 189
 China's economic resilience, 59
 China's national resilience, 59
 collective resilience, xi, 168, 183, 185–186, 191, 221–222, 225, 231
 Communities Advancing Resilience Toolkit (CART), 179, 190
 community resilience, xi, 178–181, 189–190
 and defense, 108, 127, 170–171, 182, 186–188, 198, 216, 219, 222–224, 228
 definition, 169
 individual resilience, xii, 176–178, 181
 and influence, 6, 11, 15–16, 44, 60–61, 105, 108, 169–170,

resilience
  and influence (*continued*),
    172, 176, 183, 186, 195–196,
    198–200, 202–204, 207–208,
    216, 220–221, 223, 226
  international resilience, xiii,
    181–184
  and legitimacy, 6, 11, 15–16, 44,
    60–61, 169–170, 173–174,
    176, 178, 183, 186, 195–196,
    199–200, 203–204, 207–208,
    216, 220–221, 223, 226
  national resilience, xiv, 6, 44,
    58–60, 167–169, 173–174,
    180–182, 184–185, 189–190,
    199, 208, 220, 226
  and national security, 15, 59,
    168–170, 174, 184–186,
    208–209, 216, 226
  NATO requirements, 180
  NATO's conceptualization, 182
  positive and negative role, 170
  and power, 11–13, 15–16,
    43–44, 58–61, 96–97, 108,
    127, 169–172, 176, 184, 186,
    195–196, 198–200, 202–204,
    207–208, 216, 218–221, 223,
    226, 228
  preventative measures, 170
  recommendations for grand
    strategy, 184–185
  Resilience Scale, 176, 189
  robustness and fragility, 169
  Russia's national resilience, 60
  State Resilience Index (SRI),
    58–59, 65, 176, 189, 199, 211
  state's power, influence, and
    legitimacy, 170
  supply chains, 181
  tactically and strategically
    focused, 169

resilience (*continued*)
  tools, 15, 50, 60–61, 96–97, 127,
    167–168, 170, 175–176, 179,
    181–183, 186, 195, 203–204,
    224
Responsibility to Protect, 46, 63
revisionist, 2, 8, 50, 81, 96, 120, 214
Rosenbach, Eric, 141, 162, 164–165
Russia, xi, xx, 5, 8, 27, 39, 45, 54,
  63–66, 68, 78, 80–82, 85, 87,
  100–102, 105, 107, 111–112,
  114, 129–130, 135, 146, 151, 155,
  159–161, 163–165, 206, 220–221,
  229–230
  and Austria's Freedom Party, 58
  control of information, 149–150
  debt forgiveness, 56
  disadvantage, 61
  Dozhd TV, 60
  Duma, 60
  economic statecraft, 43, 55–56
  energy dominance, 56
  fake social media, 58
  Federalnaya Sluzhba
    Bezopasnosti (FSB), 60
  foreign aid, 55–56
  Glavnoye Razvedyvatel'noye
    Upravleniye (GRU), 58
  history and geography, 47
  Internet Research Agency (IRA),
    143, 157, 162
  invasion of Ukraine, xix, 2,
    52–53, 56, 60–61, 83, 97, 108,
    123
  and Italy's Lega Party, 58
  irregular warfare, 9, 11, 43–44,
    47–48, 52–53, 55–56, 60–62,
    91, 97, 121, 123, 141, 158, 170,
    172, 201, 214–216, 224
  Kremlin Playbook, 172, 189
  law threatening journalists, 60

Russia (*continued*)
  media and non-governmental organizations, 149
  military statecraft, 3–4, 43, 52–53, 91, 104, 106, 108
  national resilience, 44, 58–60, 216
  Novaya Gazeta, 60
  nuclear weapons, 52
  paramilitary and proxy forces, 52
  Russian Cultural Centers, 58
  Russian Orthodox Church, 58, 60
  Russkiy Mir, 58
  security cooperation, 53
  Sluzhba Vneshney Razvedki (SVR), 58
  state actors and non-state proxies, 58
  state-directed and funded media outlets, 58
  System for Transfer of Financial Messages, 56
  and US elections, 58, 143, 157
  war crimes, 83, 99
  weaknesses, 61
Russian strategic culture
  and irregular warfare, 44, 47–48, 61
  pillars of, 47–48
  relative to China, 47
  relative to US, 48–49
Russo-Ukrainian War, 52, 56, 61, 108

sanctions, 3, 37, 54, 56, 60–61, 66, 77–78, 114, 119, 121–123, 125, 130, 136–137, 139, 220, 225
  financial sanctions, 124, 138
  overuse, 120, 134

Saudi Arabia, 129
Saunders, Phillip C., 51, 64–65
Schelling, Thomas, 89, 92, 110
security collaboration, 106
security cooperation, xiv, 12, 18, 51, 53, 77, 97–98, 106–108, 110, 224–225
  "advise and assist" missions, 102
  defense institution building (DIB), 102, 111
  security cooperation tools, 101–102
security coordination, 107
self-sufficiency, 108, 182–183
Shaw, Ryan, 11
shocks, xiii, 2–3, 12, 75, 118, 169, 171–172, 176–177, 181–182, 185–187, 194, 224, 226, 228
  strategic, 27–29, 34
Shyy, Jiunwei, 51, 64
Siebens, James A., 94, 110–112
silent war. *See* Kautilya
Singapore, 126
Sisson, Melanie W., 94, 110–112
Small Business Investment Company (SBIC), 135
Snyder, Jack L., 21–22, 25, 39
social media, 48, 58, 141–143, 151–152, 155, 162, 164
Society for Professional Journalists. *See* journalism
Society for Worldwide Interbank Financial Telecommunication network (SWIFT), 55–56, 65, 124, 129
soft power, 6, 13, 18, 65, 155
South Korea, 202
  Korean soldiers in US formations (KATUSAs), 101
Soviet Union, xi, xix, 1, 21–22, 39, 47, 52–53, 55, 60, 68, 70, 73–79,

Soviet Union (*continued*), 82, 85, 127, 130, 141, 149, 156, 171, 182, 195, 215, 217, 222
KGB, 48, 64
Space Development Agency, 109
spending. *See* expeditures
Stabilization Assistance Review, xix
Stalin, Joseph, 32
Star Wars, 78
State Partnership Program (SPP)
- Poland-Illinois SPP, 102–103, 111
State Resilience Index (SRI). *See* resilience, 58–59, 65, 176, 189, 199, 211
strategic competition
 legitimacy, power, and influence, xv, xvii–xix, 2–3, 5, 7–8, 10, 12–13, 15–18, 21, 29, 34–35, 37–38, 43, 45, 68, 84, 87, 91, 93–94, 97, 108, 114, 117, 121, 141–142, 145, 148, 150, 154, 161, 168, 170–172, 186–187, 193, 200–201, 203, 206, 214, 217, 225
strategic culture
 "astrategic," 32
 concept of, 23–25, 35
 comparisons of US, China, Russia, 61
 history and geography, 21–22, 27–28, 30, 44, 47
 persistence of, 27
 scholarship on, 24
 three levels, 26

strategic culture (*continued*)
 See also American strategic culture, Chinese strategic culture, and Russian strategic culture
Strategic Defense Initiative, 78
Suchman, Mark C., 14, 19
Sun Tzu, 21, 30, 35–37, 45, 63, 92, 147, 224
 Art of War, The, 39, 110, 163
supply chains, 51, 54, 60, 134, 179, 181–183
System for Transfer of Financial Messages. *See* Russia

Taiwan, 108, 119, 136, 217
Taliban, 121, 156
tariffs, 113, 118, 121–122, 137, 220
technology
 artificial intelligence (AI), 3, 59, 121, 126, 131, 135, 158, 228
 autonomous, 3, 131
 satellite, 135
terrorism, 3, 52, 91, 99, 188, 194, 219–220
 counterterrorism, xvii, 10, 102–103
threat
 cyber, 72, 83, 165, 184
 global, 2, 124, 154, 161, 169–170, 182, 186, 221–222, 228
 ideological, 27, 29
 information, 159
 most serious, 6, 44
 perception, 27, 75
 strategic competition, 2–3, 142, 148
 transnational threats, 2–6, 22, 169, 208
2022 NSS, 12–13, 79

Three Warfares. *See* China
Trade Expansion Act, 133, 137
trade liberalization, 133
Trade Promotion Authority (TPA), 133
trade war, 122, 137
Treasury Department, 124–125, 132
tripwire forces, 90, 105
Truman administration, 71, 74, 76, 107
Truman, Harry S., 71, 75–76, 107
Truman Doctrine, 74, 86
Trump administration, 7, 122, 133
trust, 58, 100, 157, 165, 189, 199, 228
　erosion of, 151, 155, 172, 174
　in government, 155, 173–174, 179–181, 202
　in sources, 145
truth, 70, 89, 137, 142, 151, 155, 160, 164, 173
　truth decay, 145, 156, 162, 167, 189, 231
Turkey, 53, 74, 107, 112, 216, 230

Ucko, David H., 17–18, 172, 189
Ukraine, xix, 2, 52–53, 60, 66, 83, 97, 99, 102, 107–108, 111, 123, 130, 137, 182, 201, 216, 220, 231
　See also Russo-Ukrainian War
United Nations (UN), 34, 79, 121, 137, 191
　International Covenant on Civil and Political Rights (ICCPR), 157
　UN General Assembly, 120, 157, 165
　UN Human Rights Council, 157, 165
　UN Office of Disaster Risk Reduction, 183
　UN Peace Operations, 52, 103

United Nations (UN) (*continued*)
　UN Security Council Resolutions, 134
United States, xviii, xx, 17–18, 28, 39, 85, 114, 137–139, 162, 164, 167, 188, 211, 230
　Cold War containment strategy, 11, 69–70, 72, 195
　competitors, 4, 9, 15, 35, 44, 54, 61, 98, 104, 107, 109, 124, 126, 130, 132, 148, 154, 156, 161, 186, 201, 206, 217–218, 228
　defense spending, 76, 109, 118, 219
　dependent on China, 68
　foreign aid, 55
　grand strategy, 1–2, 4–6, 15–16, 67, 69, 71–72, 79, 81, 84, 168, 184, 193, 213–215, 217–218, 220–221, 223, 225–227, 229
　integrated political warfare, 75, 81–82
　military culture, 5, 21, 27, 30–32, 36–37, 40, 43, 47–48, 90, 93, 223–224
　military statecraft, 3, 12, 15, 43, 51–53, 90–94, 96, 99, 103–106, 109, 126, 154, 156, 224
　nuclear umbrella, 98
　political warfare against the Soviet Union, 73
　renewed political warfare capability, 79
　reprioritization, 107, 224
　required security investments, 72
　role of global leadership, 71
　weaknesses of adversaries, 83
US Agency for Global Media (USAGM), 57, 154, 164

US Army, xx, 71
US Information Agency, 73, 154
US Information Agency (USIA), 73, 154–155
US Navy, 71
US Special Operations Command, xix
US-Morocco Free Trade Agreement, 121

Venezuela, 125
Verba, Sidney, 25, 39
Vietnam War, 76, 156, 194
violence, 10–12, 14, 18, 44, 89, 92–94, 127, 172
Voice of America (VOA), 65, 154, 161, 164–165
vulnerabilities, 46, 61, 134, 158, 167, 169, 171–172, 176, 184, 187, 223, 228
  China's increasing vulnerabilities, 217
  domestic, 218–221, 224

Wagner Group, xx, 64, 105
  Prigozhin, Yevgeny, 52
Walt, Stephen M., 230
Waltz, Kenneth N., 13, 18–19, 96, 110
war
  American view of, 34
  and peace, 9, 22, 35–36, 46, 48, 90, 92–93, 216

War Department, 71
Warsaw Pact, 47, 161
Washington, George, 31, 40, 221
weather, 2, 97, 99, 172, 178, 180, 183–185, 219, 228
Weatherford, M. Stephen, 202, 211
Williams, Heather J., 173, 189
Wilson, Woodrow, 34, 165
Wilsonianism, xv, 26, 28–29, 33–34
Wolfley, Kyle J., 64, 92, 94, 110–111
Wong, Audrye, 146, 162–163
World Bank, 113, 127, 183, 189, 198, 211
World Health Organization (WHO), 143, 162, 185, 191
World Trade Organization (WTO), 53, 113, 122, 133, 137, 139, 183
World War I, 34, 90
World War II, 28, 34–35, 39, 71, 90, 113, 127, 131
  post–World War II, 72, 214, 223
World War III, 4
Worldwide Governance Indicators. *See* indexes
Worley, D. Robert, 73, 86–87, 164–165

Xi Jinping, 44, 59, 65

Yoshihara, Toshi, 17, 78, 83, 86–87

Zimmerling, Ruth, 13, 18–19
Zolli, Andrew, 169, 188–189

About the Authors

Rebecca D. Patterson is a Professor of Practice and Associate Director of the Security Studies Program at Georgetown University and a retired US Army lieutenant colonel. Dr. Patterson served as an economist in the World Bank's Independent Evaluation Group and the deputy director of the Office of Peace Operations, Sanctions, and Counterterrorism at the State Department. She holds a PhD from The George Washington University and a BS from United States Military Academy. Dr. Patterson is a life member of the Council on Foreign Relations.

Susan Bryant is the Executive Director of Strategic Education International. She is also an adjunct professor at Georgetown University, a visiting lecturer at Johns Hopkins University and a visiting research fellow at National Defense University. Dr. Bryant is a retired army colonel whose overseas assignments include Afghanistan, Jerusalem, and South Korea. She holds a doctorate from Georgetown University as well as masters' degrees from Yale University and the Marine Corps University. Her publications include *Military Strategy in the 21st Century* and *Resourcing the National Security Enterprise*.

Jan "Ken" Gleiman is a Professor of Practice at Arizona State University's Future Security Initiative. He holds a PhD from Kansas State University, and graduate degrees from Georgetown University and the School of Advanced Military Studies. Dr. Gleiman is a retired US Army colonel who served as a Green Beret, Strategist, and was the first US Army Goodpaster Fellow. Currently the president of the Army Strategist Association and a non-resident fellow at New America, he also runs his own strategy consulting business.

Mark Troutman is an educator, consultant, and retired colonel. He is CEO of Strategic Education International and has taught business and national security economics at Georgetown University, Johns Hopkins University, University of Maryland, and George Mason University (GMU). He is the former director of GMU's Center for Infrastructure Protection and former dean of the Eisenhower School at the National Defense University. His military career included assignments in Europe, the Middle East, Asia and the US. He earned a Master of Public Policy from Harvard University, PhD (Economics) from George Mason University and is a SAMS and Army War College graduate. Dr. Troutman is the coeditor of *Resourcing the National Security Enterprise.*

CAMBRIA RAPID COMMUNICATIONS IN CONFLICT AND SECURITY SERIES

General Editor: Thomas G. Mahnken
(Founding Editor: Geoffrey R. H. Burn)

The aim of this series is to provide policy makers, practitioners, analysts, and academics with in-depth analysis of fast-moving topics that require urgent yet informed debate. Since its launch in October 2015, the RCCS series has the following book publications:

- *A New Strategy for Complex Warfare: Combined Effects in East Asia* by Thomas A. Drohan
- *US National Security: New Threats, Old Realities* by Paul R. Viotti
- *Security Forces in African States: Cases and Assessment* edited by Paul Shemella and Nicholas Tomb
- *Trust and Distrust in Sino-American Relations: Challenge and Opportunity* by Steve Chan
- *The Gathering Pacific Storm: Emerging US-China Strategic Competition in Defense Technological and Industrial Development* edited by Tai Ming Cheung and Thomas G. Mahnken
- *Military Strategy for the 21st Century: People, Connectivity, and Competition* by Charles Cleveland, Benjamin Jensen, Susan Bryant, and Arnel David
- *Ensuring National Government Stability After US Counterinsurgency Operations: The Critical Measure of Success* by Dallas E. Shaw Jr.
- *Reassessing U.S. Nuclear Strategy* by David W. Kearn, Jr.
- *Deglobalization and International Security* by T. X. Hammes
- *American Foreign Policy and National Security* by Paul R. Viotti

- *Make America First Again: Grand Strategy Analysis and the Trump Administration* by Jacob Shively
- *Learning from Russia's Recent Wars: Why, Where, and When Russia Might Strike Next* by Neal G. Jesse
- *Restoring Thucydides: Testing Familiar Lessons and Deriving New Ones* by Andrew R. Novo and Jay M. Parker
- *Net Assessment and Military Strategy: Retrospective and Prospective Essays* edited by Thomas G. Mahnken, with an introduction by Andrew W. Marshall
- *Deterrence by Denial: Theory and Practice* edited by Alex S. Wilner and Andreas Wenger
- *Negotiating the New START Treaty* by Rose Gottemoeller
- *Party, Politics, and the Post-9/11 Army* by Heidi A. Urben
- *Resourcing the National Security Enterprise: Connecting the Ends and Means of US National Security* edited by Susan Bryant and Mark Troutman
- *Subcontinent Adrift: Strategic Futures of South Asia* by Feroz Hassan Khan
- *The Next Major War: Can the US and its Allies Win Against China?* by Ross Babbage
- *Warrior Diplomats: Civil Affairs Forces on the Front Lines* edited by Arnel David, Sean Acosta, and Nicholas Krohley
- *Russia and the Changing Character of Conflict* by Tracey German
- *Planning War with a Nuclear China: US Military Strategy and Mainland Strikes* by John Speed Meyers
- *Winning Without Fighting: Irregular Warfare and Strategic Competition in the 21st Century* by Rebecca Patterson, Susan Bryant, Ken Gleiman, and Mark Troutman

For more information, see **cambriapress.com**.

Milton Keynes UK
Ingram Content Group UK Ltd.
UKHW022011290824
447530UK00007B/92